OTHER TITLES OF INTEREST FROM ST. LUCIE PRESS

THE QUALITY IMPROVEMENT HANDBOOK

TEAM GUIDE
TO TOOLS AND
TECHNIQUES

THE QUALITY IMPROVEMENT HANDBOOK

TEAM GUIDE TO TOOLS AND TECHNIQUES

Roger C. Swanson

St. Lucie Press
Delray Beach, Florida

Printed and bound in the U.S.A. Printed on acid-free paper.
10   9   8   7   6   5   4   3   2   1

Library of Congress Cataloging-in-Publication Data

Swanson, Roger C.
    The quality improvement handbook: team guide to tools and
techniques / Roger C. Swanson.
       p.   cm.
    Includes bibliographical references and index.
    ISBN 1-884015-59-X
    1. Work groups. 2. Quality control. I. Title.
  HD66.S93 1995
  658.4'036—dc20   94-46625
                                         CIP

Direct all inquiries to St. Lucie Press, Inc., 100 E. Linton Blvd., Suite 403B, Delray Beach, Florida 33483.

Phone: (407) 274-9906
Fax: (407) 274-9927

Sᴸᵗ

Published by
St. Lucie Press
100 E. Linton Blvd., Suite 403B
Delray Beach, FL 33483

DEDICATION

To my mother, Josephine,

and

To the memory of my father, Gustave

For their love, guidance, support, and trust

CONTENTS

PREFACE

The Quality Improvement Handbook had its beginning in 1985 with my Leadership Through Quality training at Xerox. After this training, I participated in numerous quality teams where I realized the benefit of quality tools and techniques. These simple but effective methods of logically and systematically attacking problems enabled teams to quickly define, understand, analyze, and track process improvement activities.

In 1988, I was fortunate to lead a team that won a Team Excellence award in the Special Markets Group at Xerox. Our team, the SMURFS (Special Markets Underutilized, Redundant Financial Systems), eliminated a financial reporting system that paralleled the corporate system. The benefit of teams using quality tools was obvious. Later that year, I developed a two-hour training module to promote benchmarking awareness for managers. I was following the Xerox concept of LUTI, learn, use, teach, and inspect for quality.

My background at Xerox and my interest in quality management led to the formation of my consulting firm, Competitive Dynamics, Inc., in 1990. Our mission is to provide quality management training and consulting in both the public and private sectors. Industries served include agriculture, chemical, government, healthcare, military, office products, petroleum, printing, and utilities—evidence that the benefit of using quality management tools and techniques transcends industry boundaries.

The next milestone that led to this book was my decision to become involved in the UCLA Extension Total Quality Management certificate program. This program had just begun and there was a need for additional courses. After contacting Ken Farrell, who was still at Xerox at the time, and one of the TQM curriculum advisors, I agreed to develop the course in "tools and processes." This course is one of five "core" courses and is titled "Introductory Quality Tools and Processes." Together with a former co-worker at Xerox, Jules Cochoit, the course material was developed and presented in the fall of 1991.

Our first textbooks were Xerox training books on problem solving, process improvement, and benchmarking, plus two other textbooks. When we added the seven "new" management tools to the curriculum, we added another book. Total required reading exceeded 650 pages! Several tools and techniques were covered in great depth, well beyond the needs of most quality teams. Other tools received less than adequate coverage and some were not presented at all. *The Quality Improvement Handbook* emerged out of this need to present truly useful tools and techniques without burdening students with information not necessary for them to lead teams.

I wrote this book to provide quality improvement teams, and their leaders, with a comprehensive set of tools and techniques to solve problems and improve processes in their organizations. Included are detailed descriptions and instructions for the original seven quality tools, the seven "new" management tools, and more than twenty additional tools and techniques. The book uses an eight-step Quality Improvement Model that provides a logical framework for applying each tool and technique. The model is presented in summary form for quick reference, as well as in a detailed descriptive format for teams that need more than the summary version. It provides a level of detail frequently needed to guide teams but often lacking in other models. The book includes a glossary that helps to summarize the key concepts presented.

Each tool and technique is presented in a standard format that first describes it and then identifies key points and typical applications. This is then followed by an example and detailed steps to use the tool or technique. For those readers who need more information, a listing of other resources and references is presented in Appendix B.

Roger C. Swanson

THE AUTHOR

Roger C. Swanson is an instructor and curriculum advisory committee member for the University of California, Los Angeles Extension Total Quality Management certificate program. He graduated from the University of California, Berkeley with a B.S. in Business Administration and received his Master's degree in Business Administration from the University of California, Los Angeles.

Mr. Swanson is President of Competitive Dynamics, Inc. and has over 20 years experience with Fortune 100 companies (Xerox, Unocal, Alcoa, and Procter & Gamble). His career includes both line and staff positions in the electronics, office products, consumer products, packaging, and chemical industries. He has developed expertise in benchmarking, competitive analysis, finance, and logistics in support of manufacturing, engineering, and marketing activities. Mr. Swanson has a proven track record of initiating and implementing projects to improve operational performance in diverse industries.

Mr. Swanson is a regular speaker at professional organizations in Southern California. He has presented numerous workshops on benchmarking, quality management, and quality function deployment, including workshops for the American Society for Training and Development Quality Symposia in 1993 and 1994. He has been an active member of the American Society for Quality Control, the American Society for Training and Development, the Council of Logistics Management, and the Society of Competitive Intelligence Professionals.

Mr. Swanson has published articles in *Quality Progress* and the *Journal of Applied Manufacturing Systems*. He also contributed the chapter on benchmarking in the Society of Manufacturing Engineers' *Continuous Improvement Handbook*.

ACKNOWLEDGMENTS

Several people helped in the development and refinement of this book. Their invaluable comments, critiques, and suggestions helped immeasurably.

First, I would like to thank Eric Johnson for his sage thoughts and constructive criticism. His in-depth suggestions and analysis influenced the major revisions from the original draft.

I would also like to thank the entire "Introduction to Quality Tools and Processes" class (UCLA Extension, TQM Certificate Program, fall 1993) for their patience, advice, and constructive criticism. I would like to thank in particular Gayle Torres, Kevin Clark, and Blaine Wiseman, Jr. for taking the time to offer detailed comments and provide a detailed rating of the book topics and format. A special thanks to my co-instructor in the "Tools and Processes" class, Jules Cochoit III, for his thoughts and constructive comments.

To the many friends and colleagues who offered encouragement during this endeavor, I offer a special thank you.

CHAPTER ONE

INTRODUCTION

QUALITY AND ITS PREREQUISITES

Although there are numerous definitions of quality, at least one for each "guru," this book accepts the International Organization for Standardization (ISO) definition as a starting point: Quality is the totality of features and characteristics of a product or service that bear on its ability to satisfy given needs. Implicit in this assumption is a customer with needs and expectations to be satisfied. By adding a time dimension, we further assume that quality is not static: standards of performance increase with time.

Quality is a measurable result, but, as we will soon see, quality improvement is more about the journey than the end result. What we do and how we do it determine quality; thus, the more interesting and actionable elements of quality involve the process rather than the outcome.

The key ingredient to begin the journey is senior management's commitment to openly and genuinely embrace the concepts, be willing to change, provide the resources, and create the environment in which the change will be successful. It is a change in attitude that must result in changes in behavior. For many organizations, there are barriers to changing the culture to accept quality improvement; these barriers can include past success without quality management, a strong leader or a bureaucracy that refuses to change, and a technology based on "creative genius" rather than teamwork.

Once the first hurdle has been cleared, quality management can be described in terms of several key characteristics:

- Customer focused
- Decisions based on facts, statistical data, and knowledge

1

- Strategic, systematic continuous performance improvement involving everyone
- Focused on outcomes and results
- Teamwork performed by empowered employees who:
  - Are well trained
  - Understand goals and objectives
  - Are committed and take responsibility
  - Operate within assigned scope or boundaries
  - Are stakeholders in the process
- Long-term focused
- Addresses key cross-functional processes, not just products or services
- Prevention, not inspection
- Based on commonly held organizational values (as expressed in the vision, mission statement, goals, and objectives of the organization) deployed to operational work units
- Open communication to build trust and teamwork
- Leaders who motivate, not managers who direct
- Elimination of departmental barriers
- Team-based recognition system

This organizational change does not just happen once senior management makes its commitment. The organization must plan an implementation that includes training, a system to track progress, a reward and recognition system, a customer satisfaction measurement system, etc. Senior management may also form a Quality Council to establish the plan and a Steering Committee to represent functional departments in guiding the implementation. These organizations can be extremely helpful in assisting the smooth implementation of quality improvement.

Finally, training must be delivered to provide employees with the methodologies to improve quality. To be effective, this training should be immediately applied to problems and improvement opportunities in the work environment. This book is designed to assist in such "just-in-time" training.

ABOUT THIS BOOK

The purpose of this book is to provide quality improvement teams with a comprehensive set of tools and techniques to solve problems and improve processes in their organizations. Included are detailed descriptions and instructions for the original seven quality tools, the seven "new" management tools, and more than twenty additional tools and techniques.

The book uses an eight-step Quality Improvement Model that provides a logical framework for applying each tool and technique. The model is presented in summary form for quick reference, as well as in a detailed descriptive format for teams that need more than the summary version. It provides a level of detail frequently needed to guide teams but often lacking in other models.

The first three chapters provide a summary of the book and background ideas necessary to use the Quality Improvement Model presented in Chapter Four. The primary tools and techniques used by quality improvement teams are discussed in Chapters Five through Twelve. The examples used throughout the book follow a case study entitled Video One, which is presented in Chapter Fifteen. It is recommended that the case study be read after completing Chapter One.

Chapter Five: Idea Generation

Chapter Six: Consensus

Chapter Seven: Process Definition

Chapter Eight: Collecting Data

Chapter Nine: Analyzing Cause and Effect

Chapter Ten: Analyzing and Displaying Data

Chapter Eleven: Planning Tools

Chapter Twelve: Meeting Management Tools

Chapters Thirteen and Fourteen present special tools used less frequently: Benchmarking and Questionnaires.

The names of tools and techniques are referenced as capitalized proper nouns throughout the text. A synopsis of the fifteen chapters follows.

Chapter One: Introduction

An overview of the entire book is provided in this chapter. It begins with a brief discussion of quality management and the prerequisites that enable quality team success.

Chapter Two: Work as a Process

This chapter establishes the foundation for several key principles upon which the book is based. It sets the stage for the discussion of viewing an organization in terms of its processes rather than its functional departments. Organizations need to change from the old functional structure to an organization composed of a series of processes that cross old functional lines. The functional expertise of the organization is retained, but workers interact differently.

The changes described not only change the organizational structure, but also require a cultural change. These "structural" factors (i.e., organization structure, culture, environment, and technology), along with the processes and practices of the organization, drive performance. Structural factors, however, become barriers to performance that must be either overcome or accepted before the quality journey can begin. The focus of quality improvement then becomes process improvement.

One key element of quality management is the dependence on empowered teams to document and then streamline processes based on team knowledge and understanding of customer needs and expectations. (Stated customer requirements often fall short of defining customer needs and expectations. Therefore, the latter terminology is used to ensure that teams first understand needs and expectations before defining requirements.)

By focusing on the customer, the organization uses customer needs and expectations to guide it in "doing the right things" or using judgment. This judgment selects and modifies the processes, practices, and structural factors that drive performance. In order for the organization to be effective in producing desirable outcomes, it will have to execute efficiently, that is, "do things right." An understanding of these concepts will help teams select quality improvement projects that are important to the organization and over which they have control.

This chapter reviews the first and often the most critical issue faced by quality teams: identifying and selecting important problems to solve or processes to improve. This selection effort is influenced by strategic and competitive assessments and goal alignment which guide teams in making the right choices. Finally, a technique to assist work unit teams select a process to improve is presented.

Chapter Three: Measures of Effectiveness

Quality improvement requires knowledge of how well customer needs and expectations are satisfied. This chapter begins with a discussion of the importance of identifying and measuring the vital few desirable outcomes and related process measures. These data form a family of measures that becomes the basis for quality improvement decisions. Data collection occurs at several points, beginning with data available at project inception, followed by collection of baseline data to verify the existence of the problem or opportunity, and finally by data collected to verify root causes and monitor implementation of solutions.

Quality improvement is more than collecting data, however. It requires an understanding of the processes that produce products and services and the link to customer needs and expectations. Desirable outcomes measure organization effectiveness in satisfying customers. More importantly, teams must understand the causes of desirable outcomes if they are to improve them. This cause-and-effect relationship is at the heart of quality improvement. The challenge is to gather and analyze the vital few measures and to determine the drivers of performance that cause the desired effect or outcome. Success depends on changing root causes—the drivers—instead of attempting to change secondary causes or symptoms.

This chapter briefly discusses the dimensions of quality used to measure customer satisfaction. These dimensions, along with financial, time, and quantity-related measures, are the general categories of performance indicators used in the family of measures.

Finally, a review of data types helps teams understand the terminology used to describe various tools and techniques presented. The chapter ends with a discussion of data validity, reliability, and bias.

Chapter Four: Quality Improvement Model

The eight-step model presented here is a generic model that follows the Plan-Do-Study-Act cycle. Step 1 of the model guides the team in selecting a problem/process improvement opportunity. Step 2 leads the team through the important activities needed to define customers, outputs, customer needs and expectations, supplier specifications, and baseline data. In this step, teams analyze the current situation by documenting processes and practices. In Step 3, the search for root causes is conducted, and potential solutions are evaluated and selected in Step 4. Step 4 also addresses detailed implementation planning activities, including countermeasures to offset unexpected events during implementation.

Steps 5 through 7 follow the standard organization change process of first implementing pilot solutions followed by monitoring of results and then studying the solution. This then leads to development of training materials, if required, to facilitate the standardization of change throughout the organization in Step 7. Finally, Step 8 completes the cycle and leads back to Step 1 and the next cycle of continuous improvement.

Chapter Five: Idea Generation

This chapter describes tools used extensively during the Planning phase of quality improvement to explore opportunities for improvement and potential solutions. Teams use these tools to capture and benefit from the expertise and diversity of the team members in an effort to expand ideas prior to attempts to reach consensus.

The idea generation tools and techniques include:

• Brainstorming

• Brainwriting

• Crawford Slip Method

Chapter Six: Consensus

This chapter describes tools used extensively during the Planning phase to help teams select opportunities or solutions for which full agreement and support are needed. These tools help teams reach consensus, which goes beyond the voting elements of some of the tools. Voting helps to test the pulse of the team, but does not result in decisions or consensus. (The original seven quality tools are denoted by an asterisk [*] and the seven "new" management tools by a double asterisk [**].)

The consensus building tools and techniques are:

- Affinity Diagram**
- Balance Sheet
- Criteria Rating Form
- Is/Is Not Stratification
- List Reduction
- Paired Comparisons
- Weighted Voting

Chapters Seven through Ten: Analytical Tools and Techniques

Chapters Seven through Ten contain the primary analytical tools and techniques used in quality improvement. The tools are categorized according to their primary use. These tools and techniques guide teams in collecting, analyzing, and displaying data to help them understand relationships, particularly cause-and-effect relationships, and to define processes. (The original seven quality tools are denoted by an asterisk [*] and the seven "new" management tools by a double asterisk [**].)

Chapter Seven: Process Definition

- Flowcharts
- Process Analysis Worksheet
- Tree Diagram**

Chapter Eight: Collecting Data

- Check Sheets*
- Focus Groups
- Sampling
- Surveys

Chapter Nine: Analyzing Cause and Effect

- Cause-and-Effect Diagram*
- Five Whys
- Interrelationship Digraph**

Chapter Ten: Analyzing and Displaying Data

- Charts (Bar, Pie, Run, and Spider)*

- Force Field Analysis
- Histogram*
- Matrix Diagram**
- Pareto Chart*
- Prioritization Matrix**
- Scatter Diagram*
- Statistical Process Control: Control Charts*
- Statistical Process Control: Process Capability

Chapter Eleven: Planning Tools

These tools and techniques aid teams in planning and tracking the implementation of solutions. Storyboards, Tree Diagrams, and Gantt Charts are quick, easy-to-use tools to plan and document project activities. In addition, Storyboards are particularly useful in communicating and documenting quality improvement activities, such as presentations to management, and communicating progress in newsletters or on public bulletin boards. (The original seven quality tools are denoted by an asterisk [*] and the seven "new" management tools by a double asterisk [**].)

- Activity Network Diagrams**
- Gantt Chart
- Process Decision Program Chart**
- Storyboards
- Tree Diagram**

Chapter Twelve: Meeting Management Tools

Meeting management tools aid teams in improving meeting effectiveness and in documenting meeting actions, decisions, and discussions. They help teams manage both the process and content of meetings.

- Agendas/Meeting Minutes
- Plus/Delta (+/Δ) Evaluation

Chapter Thirteen: Benchmarking

Benchmarking is an externally focused quality improvement technique that helps teams discover how other organizations are performing functionally equivalent activities. This type of Benchmarking is often referred to as operational or functional Benchmarking. It often follows some form of competitive analysis and can be used as a stand-alone process or in support of internal improvement efforts. This search for best practices provides teams with a different view during quality improvement that is often the spark for "creative imitation," or the modification of innovative methods to one's own organization. In other situations, Benchmarking helps to validate root causes and proposals for change.

Chapter Fourteen: Questionnaires

Questionnaires are useful in collecting data during various stages of quality improvement and are an integral part of the primary field research required for Benchmarking. A well-designed questionnaire is invaluable in collecting valid, reliable, unbiased data. Uses of questionnaires vary with the information sought and the population being queried. Depending on time limits, number of people questioned, length of the questionnaire, and level of detail, questionnaires are designed for either written or verbal (either in person or by telephone) responses.

Chapter Fifteen: Video One Case Study

The examples used throughout this book are based on the Video One Case Study. Although the instructions for tools and techniques are self-explanatory, the examples are generally easier to comprehend when the reader understands the context of their use. Therefore, it is recommended that Chapter Fifteen be read **before** reading Chapters Five through Twelve.

Appendices

The appendices include a glossary of terminology, references that were used in developing this book, and additional resources that readers can use to expand their knowledge of quality improvement.

   A: Glossary

   B: References and Resources

WHAT THIS BOOK IS NOT

Much of Total Quality Management is not addressed in this book. Because the focus is on tools and techniques used by work unit teams in both manufacturing and non-manufacturing organizations, specific subjects were excluded. These subjects are, nonetheless, important to some readers, and several excellent resources are identified in Appendix B (References and Resources).

   In order to be effective, teams need to understand the "human side" of quality, which was excluded from this presentation. Teamwork, leadership, and communication skills are key to the success of quality teams, and other resource materials should be reviewed for guidance in group interaction skills, team roles and responsibilities, and other elements of group dynamics. Teams also need to understand the characteristics of leadership and how it affects employee involvement and ultimately empowerment.

The primary objective of the book is to assist work unit teams to improve their processes. As a result, specific tools to plan a quality management system, to assess organizational readiness to implement quality management, or to deploy organization policies or themes were excluded. Many of the tools presented are applicable to these tasks, but no model, such as the Quality Improvement Model, is included for these important activities.

Finally, several tools were excluded because a primary criterion for inclusion was applicability to both the manufacturing and non-manufacturing environments. For example, tools used primarily in manufacturing and engineering environments, such as design of experiments, probability plots, statistical process control for short and small runs, and failure mode and effects analysis, were excluded. Again, excellent resources are listed in Appendix B for readers seeking additional information.

CAUTION ON "TOOLBOX TQM"

The objective of using the Quality Improvement Model with various tools and techniques is to solve problems and improve processes. The model is only a guideline; it is not a recipe for success. The reader should exercise judgment in using it. Flexibility in applying the tools is also helpful, because each tool has strengths and weaknesses. The tool should fit the problem or situation, not the other way around. Often, there is no one right way. A variety of analytical tools need to be applied, along with group dynamic and interactive behavior skills that help teams become effective. Do not fall into the trap of "toolbox TQM," where the problem or issue is secondary to the tool and where the group dynamics are ignored.

Why is this caution added here? Some teams fall into the trap of using the same tools repeatedly, often inappropriately. They become a slave to a tool! (The practical applications of Control Charts, for example, are often more limiting than their theoretical applications.) Do not let the means to an end become an end itself that becomes another non-value step. A tool is not a solution in search of a problem to solve!

CHAPTER TWO

WORK AS A PROCESS

WORK: A SERIES OF ACTIVITIES

All work is a process. It is a series of activities that produces products and services whereby inputs are transformed into outputs following a sequence or pattern of events performed by a variety of individuals.

In traditional organizations, this work was generally arranged around functional activities, which often suboptimized performance because of allegiance to individual departments. In many cases, the voice of the customer, either internal or external, was never heard. If minimum specifications were met by a department, the work was complete. If there were problems, inspection would catch them. Final product and service performance measures were not identifiable at the work unit level. The process lacked documentation, except perhaps at the departmental level, and this lack of a system resulted in significant performance variations. Frequently, the end result was seriously flawed products and services.

Total Quality Management is about solving these problems, which are perhaps more accurately defined as process problems: either a repeatable, defined process is absent or a poorly designed process exists. Over the years, this attack on poor performance has been conducted under the banners of systems analysis, productivity improvement, and, now, re-engineering. The words change, but the desired outcome is the same: process improvement.

Today, however, the people performing the change are likely to be stakeholders or owners of the process—those individuals who perform the work—rather than some staff group. Instead of individuals solving problems, we now have teams. They understand the process or suffer in the absence of one; they know what practices are used to perform their work. These practices determine how activities within a process are performed. They range, for example, from manual versus automated activities to communication using written reports versus oral presentations in meetings. In short, the processes and practices drive performance (often referred to as performance drivers)—and who knows this better than the worker performing the activities?

For many readers, the term *process improvement* means continuous yet small improvements in a process. That definition is expanded here to include more radical re-design, or re-engineering, of processes. Radical changes, or breakthroughs, that result from the efforts of empowered teams using tools such as Deployment Flowcharts are part of process improvement. For example, a team might recognize that communication and computer technology could enable the combining of functional departments. The point is that a team that includes stakeholders and owners of a process can improve the process in either small or large increments of change.

In the near term, there will be much conflict as middle managers cling to the old functional structure from which they derived their power. Those managers with functional expertise who focus on cross-functional processes will survive in the long term. Their power will come from their expertise and how effectively they can work with others as a group to satisfy customer needs and expectations.

The benefit of having teams perform the work more than offsets the need to train them to work together. In addition, if teams are armed with some fairly simple tools to harness their collective knowledge and efforts, the benefits are enhanced. That is what the Quality Improvement Model and the tools and techniques described in this book are intended to do.

Not all teams are the same, and group dynamics vary between teams composed of members from the same functional or operational work unit or teams that are cross-functional. In either case, the tools and techniques presented here can be applied uniformly with equal success.

FUNCTIONAL VERSUS PROCESS ORIENTATION

Our quality journey begins with the notion that work is performed as a cross-functional process rather than a series of functional activities performed within discrete departments. Individual functional or operational departments in an organization seldom produce a complete product or service without interacting with others in the organization. Eliminating departmental barriers takes time, however, even after the organization has accepted the process orientation. Work follows a torturous path as it moves from department to department (see Figure 2.1).

As the organization changes to accommodate the shift in focus from well-defined departmental "silos" to cross-functional processes, it becomes more flexible. Departmental walls become invisible as the process and its outputs become more visible. Continuous improvement and documentation (with Flowcharts and baseline performance data) result in processes that vary less, take less time, have fewer steps which all add value, and are more robust (i.e., simpler,

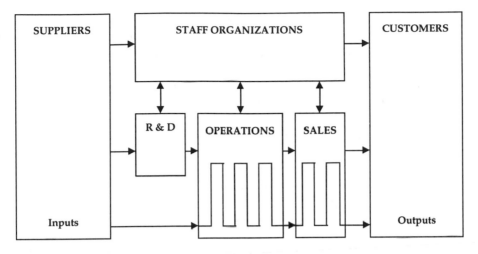

FIGURE 2.1   Traditional Company

more flexible, and mistake-proof). Concurrent or parallel activities replace sequential actions and missed hand-offs as work units become organized around key processes (see Figure 2.2).

The shift from the traditional to the process orientation is enabled by focusing on customers. When the organization focuses on customers, it will recognize that the process is a series of customers, each served by a supplier that adds value. The process is a string of outputs of products and services that depend on outputs of previous steps. Each step should add value, and together they become a chain of activities that create value—a process.

The process begins with external suppliers, progresses through the organization in a series of internal supplier and customer relationships, and ends with

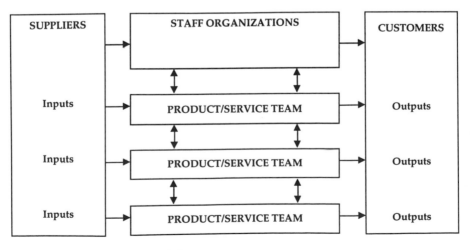

FIGURE 2.2   Process-Oriented Company

external customers. Addressing these customer and supplier relationships hastens the transition to a process orientation. The negotiations and resulting quantifiable measures that document this relationship are explored in Chapter Three.

In addition to becoming more robust, processes become more efficient (i.e., greater output per unit of input) and effective (i.e., more desirable outcomes) with each improvement. Rather then setting procedures in stone, the goal is to think about what can go wrong and be prepared with flexible countermeasures. Performance becomes insensitive to uncontrollable variations due to internal and external factors. There is a price to be paid for this robustness in the form of added costs, but with the unknown and unknowable costs of process variation, these added costs are usually justified.

The use of creative problem solving often minimizes this cost of robustness because it is focused on simplifying processes, reducing human errors and non-value steps, looking at early and important steps in the process, and capturing additional data where failures occur. Creativity can add robustness without significant cost.

BARRIERS TO CHANGE: STRUCTURAL FACTORS

Quality teams needs to address a special category of performance drivers: structural factors. These are the culture, organization structure, technology, environment, and any other administered factors (such as contractual agreements) that often influence performance. Structural factors create barriers that must either be corrected or accepted as a given but should never be ignored. For example, an organization with too many layers, the wrong structure, too many job categories, and/or a culture that is not customer focused needs to understand the effect of these conditions.

Generally, structural factors must be addressed first to prevent disruptions during implementation of the quality program. Assessment of organization readiness to implement quality management concepts uncovers these barriers. Management's commitment to quality principles begins the culture change, which is often the most significant barrier.

In most instances, structural factors are not within the control of work unit teams and must, therefore, be accepted as a given. Organizations that are faced with financial pressures due to an inefficient organization structure, for example, often need to restructure before applying quality management principles. To do otherwise will likely create morale issues that detract teams from their improvement efforts. If radical changes are required, senior management needs to re-engineer before empowering work units. Once work units have experience with quality improvement, they too will be capable of re-engineering efforts.

PROCESS CLASSIFICATIONS

Systematic classifications of processes have just begun to emerge out of the quality movement, particularly from organizations doing Benchmarking. Their incentive for developing a classification system is to facilitate the exchange of data about similar processes. Although no universally accepted scheme has been developed, several versions separate processes into operating versus management, or support, processes. Management processes focus on activities to:

- Manage resources (human, financial, information, and physical)
- Deploy policies (i.e., align operational objectives with the organization's goals and objectives) and manage organizational change
- Manage external relationships other than with customers

The operational processes include:

- Market and customer research
- Product and service design
- Marketing and sales
- Production and delivery
- Customer service

PROCESS IDENTIFICATION AND SELECTION

For most work unit teams beginning the quality journey, the first problem encountered is the identification and selection of a process to improve. Many organizations use Steering Committees and Quality Councils to guide teams in this critical stage. Two key organizational activities facilitate the work unit agenda: competitive and strategic assessments and goal alignment (often called step zero or policy deployment). Strategic and competitive assessments provide a framework that guides the organization, and goal alignment links work unit objectives with the organization's goals and objectives (see Figure 2.3).

One element of a competitive and strategic assessment should focus on the customer (and obviously competitors and strategic financial, marketing, and technological issues) in an attempt to gain knowledge about customer needs and expectations. The objective is to identify what the organization must focus on in order to satisfy customer needs and expectations. (Critical success factors is the term often used to describe what the organization must address in order to be successful. This term is used sparingly here because it is often confusing. Instead, important customer needs and expectations define what is critical to the success of any organization.)

Once the organization conducts its assessment and understands what is critical to its success, it is then ready to define its vision and its strategic direction. The vision is consistent with and builds on the organization's mission statement, which is a statement of goals that defines the organization's purpose, its reason for existing, and its philosophy. This philosophy explains how the organization values and utilizes its resources.

With the vision defined, the organization next addresses its goals and objectives and finally its strategy for achieving the vision. This effort usually results in two documents: a long-range plan (three to five years) and an annual plan. The judgment portion of Figure 2.3 illustrates these basic activities. These plans not only establish specific goals and measurable objectives—the desired outcomes— but must also address the means to reach these outcomes.

Reaching desired outcomes occurs when the organization correctly identifies the cause-and-effect relationships between its strategy (the "how to" or means) and its goals and objectives (the "what"). The cascade of goals and objectives at one level as the means to achieving those goals at the next level below is at the heart of policy deployment (see Appendix B: References and Resources, General Reference). This is the execution portion of Figure 2.3.

The organization strategy often requires definition of the performance drivers (the processes, practices, and structural factors) needed to achieve goals and objectives. Unfortunately, performance drivers at the organization level are a summary view of the detail required at the work unit level. When policy deployment is absent, which is typical, work unit teams often have difficulty linking

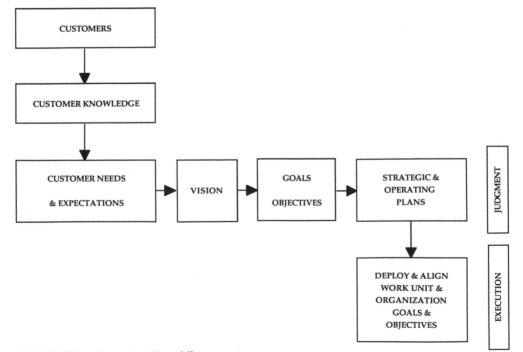

FIGURE 2.3 Organization Alignment

their activities to specific organizational processes. This dilemma is addressed in the next section (Process Identification for Work Units).

Management's judgment in selecting strategies consistent with customer needs and expectations becomes a prerequisite for organization effectiveness. This judgment, or "doing the right thing," completes the first part of the effectiveness equation.

Next, the organization must execute, that is, "do things right." Hence, effectiveness depends on both judgment and execution. At the organization level, execution involves the cascade of goals and objectives to all levels in the organization. This deployment and the resulting goal alignment define the execution portion of organization effectiveness. (Policy deployment is not management by objectives or some other top-down targeting methodology. Instead, it involves the cascade of negotiated objectives ["what"] and the definition of the means to achieve those objectives ["how to"]. It requires communication, negotiation, and commitment at all levels.)

If the organization is effective in selecting the right goals and objectives and if it deploys its strategy properly, it will be aligned with customer needs and expectations. Next, work units must use judgment and execute properly to ensure that they are effective in producing the products and services needed and expected by customers (see Figure 2.4).

Customer knowledge from Surveys, suggestions, complaints, and Focus Groups helps to define in detail customer needs and expectations at the work unit level. Translating these needs and expectations ("what") into the means for achieving them—the performance drivers ("how to")—reflects judgment at the work unit level.

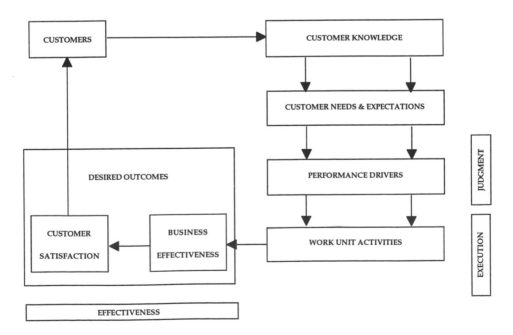

FIGURE 2.4 Work Unit Alignment

A process defined at the organization level, for example, has many sub-processes defined at the work unit level. Understanding the organization's goals and objectives through policy deployment provides the basis for work unit judgment. Next, the work unit must execute and measure its performance.

Work unit performance measurement that addresses both supplier and customer relationships as well as the achievement of organization objectives is discussed in Chapter Three. The primary objectives, or outcomes, are customer satisfaction, customer loyalty, and business effectiveness. Measures not linked directly with customers are part of the latter category, which includes such measures as return on assets, profitability, and market share. This is represented in the effectiveness portion of Figure 2.4.

A simple approach to guide work units in selecting processes to improve is explored in the next section. Unfortunately, most organizations have not aligned goals and identified key processes that are linked to desired outcomes as described above. Furthermore, work units often have difficulty in associating activities with specific processes. This technique helps to bridge that gap.

For work units linked with external customers or those that understand internal customer and supplier relationships, an alternate approach is discussed in Chapter Four. This alternative uses a Process Analysis Worksheet that starts with the customer (in Step 2 of the Quality Improvement Model). It begins with the output received by the customer and chains backward through the process to the work unit involved and, if needed, to internal and external suppliers serving the work unit.

PROCESS IDENTIFICATION FOR WORK UNITS

For most work units, the policy deployment approach of selecting a process (described above) is at best a wish rather than a reality. Their question is more basic: How does a quality team identify a process to improve?

One solution (see Figures 2.5 and 2.6) uses Matrix Diagrams (Chapter Ten: Analyzing and Displaying Data) to first identify important activities and then link these activities with process categories. The items selected for the axes of the matrices are developed by the team using Brainstorming (Chapter Five: Idea Generation).

The following example relates to the Video One Case Study presented in Chapter Fifteen. First, read the sections on Industry Background and Selecting an Improvement Opportunity in Chapter Fifteen for background on the example used here.

1. In the first step (Figure 2.5), teams compare major activities performed (the vertical axis, left side) with the key goals and objectives of the organization (horizontal axis, at the top). (This, of course, assumes that organization goals and objectives have been communicated, which is a prerequisite of quality improvement.) The team then assesses the relationship between its activities and goals/objectives. Where an activity

ACTIVITIES	GOALS/OBJECTIVES				
	Satisfied Customers	Inventory Control	Attract New Customers	Increased Sales	TOTAL WEIGHTED POINTS
WEIGHTING	X 4	X 1	X 2	X 3	
Order New Tapes	9	1	3	3	52
Prepare Customer Survey	3	0	0	0	12
Analyze Customer Survey	9	0	3	3	51
Prepare Advertisement	3	0	9	3	39
Place Advertisement	3	0	9	9	57
Stock Display Shelves (Empty Boxes)	3	3	0	0	15
Update Manual Inventory Records	1	9	0	1	16
Sales Transactions					
Deliver Tape to Customer	3	3	0	1	18
Complete Credit Card Transaction	3	0	0	0	12
Write Manual Invoice	3	1	0	0	13

FIGURE 2.5   Important Activities

supports, or is related to, a goal/objective, a value is placed in the intersecting box: 0 = no relationship, 1 = possible relationship, 3 = medium relationship, and 9 = strong relationship. In Figure 2.5, we see that the goal of Satisfying Customers is strongly related to the activities of Ordering Tapes and Analyzing Customer Surveys and has some relationship to all other activities, except the Updating of Manual Inventory Records.

Goals and objectives can be weighted to reflect their relative importance. For example, in Figure 2.5 the goal of Inventory Control is given a weighting value of 1 and Increased Sales a value of 3. This means that Increased Sales is three times as important as Inventory Control.

Next, sum the values for each row, including the weighting of goals/ objectives times the value in each box of the row. For example, Ordering New Tapes has a value of 52 (i.e., $9 \times 4 + 1 \times 1 + 3 \times 2 + 3 \times 3$). Activities with the highest row values are important activities for the work unit.

2.   Next, the team compares important activities (horizontal axis) with processes (vertical axis). A good starting point for listing processes is presented in the Process Classification section earlier in this chapter. Add additional processes and subprocesses appropriate to the work unit. Use the technique in step 1 to assess the strength of the relationships between activities and processes. In Figure 2.6, we see that the Advertising subprocess, which is part of a more general Marketing and Sales process, is strongly related to Preparing and Placing Advertisements but has some relationship to Ordering New Tapes and Analyzing surveys. Sum the values by row to determine important processes.

| | IMPORTANT ACTIVITIES | | | | | |
PROCESSES	Order New Tapes	Analyze Surveys	Prepare Ads	Place Ads	Sales Transaction	TOTAL POINTS
Human Resource Management	0	0	0	0	0	0
Inventory Management	3	0	0	0	3	6
External Relations	0	0	0	3	0	3
Market & Customer Research						
Survey Design & Analysis	1	9	3	0	0	13
Marketing & Sales						
Advertising	1	1	9	9	0	20
Check-Out	0	0	0	0	9	9
Production & Delivery						
Title Selection & Ordering	9	3	0	0	0	12

FIGURE 2.6   Important Processes

3.   Finally, the team must select important processes that need improvement. This rating of performance can range from a self-assessment to a survey of customers who use the output (a product or service) of the process. In Figure 2.7, we see that the Advertising process is the most important process, and it appears to be effectively performed: it is one of the strengths of Video One. Both the Check-Out process and the Title Selection and Ordering process are also important but appear to be less effectively performed. These are opportunities for improvement.

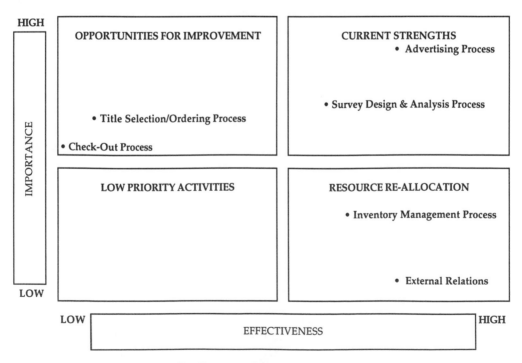

FIGURE 2.7   Importance/Effectiveness Matrix

This technique is primarily geared toward teams from the same work unit or department. Teams representing various functional departments (i.e., cross-functional) can apply the technique, however, on either a department-by-department basis or as a single cross-functional team.

As teams gain experience, they will likely employ other quality tools to assist them in selecting opportunities for improvement. After reviewing the next chapter on Measures of Effectiveness, the team will be ready to apply the Quality Improvement Model to improve selected processes.

CHAPTER THREE

MEASURES OF EFFECTIVENESS

The primary objective of organizations is to satisfy customers. Organizations that consistently produce products and services that satisfy customers are the ones that effectively serve the needs and expectations of their customers. Effectiveness measures the organization's ability to achieve desired outcomes: Were customers satisfied? Were our customers loyal? Did market share and profits increase? Were new products introduced before competitors' products?

Outcomes are not outputs. We can productively produce a multitude of products and services (outputs) and satisfy no one (an undesirable outcome). For our purposes, productivity is defined as efficiency (i.e., output per unit of input) times utilization (i.e., actual use of system versus theoretical use). Hence, productivity does not guarantee effectiveness, which adds the dimension of satisfying customer needs and expectations. Productivity and effectiveness are treated separately in this book.

These desirable outcomes, or results, can be measured in a variety of categories. We will focus on only the vital few outcome measures that define organization effectiveness. Outcomes, along with process measures and attributes, form a family of measures that are explored in this chapter. Measurement begins with baseline data developed in the early phases of quality improvement and continues through implementation.

Decisions based on facts are a key element of quality improvement, but data alone is not enough. The organization needs to understand how it achieved its results, that is, to understand what drives performance that satisfies customer needs and expectations. The performance drivers (i.e., the processes, practices, and structural factors) introduced earlier must be changed if results are to improve. This cause-and-effect relationship between performance drivers and outcomes is at the heart of quality improvement. Our goal, then, is to link outcomes with the processes and practices that produced them and with structural factors that influenced them.

THE EFFECTIVENESS EQUATION

Measurement of results, or outcomes, is a difficult yet necessary task in determining organization effectiveness. The difficulty arises because effectiveness depends on selecting the "right things to do" and "doing things right," or judgment and execution. Judgment creates the system of performance drivers, while execution determines how productively the system performs (i.e., efficiency, or output per unit of input, times utilization of the system). Poor results usually occur when either or both of these elements fail. Superior execution can seldom overcome wrong or poorly chosen drivers. Although quality improvement addresses both elements of effectiveness, the search for key causes is simplified if the failure can be identified as one of either judgment or execution.

As would be expected, unfavorable outcomes often mask excellent processes that are executed poorly. This is the halo effect, where an organization that performs well is often assumed to do everything well, and conversely, an organization that performs poorly is assumed to do everything poorly. In most quality improvement activities (and particularly in Benchmarking), organizations with poor outcomes are generally ignored because of credibility issues.

Quality improvement occurs only when performance drivers are changed or executed properly. No amount of target setting, or wishful thinking, that addresses only process measures and outputs will have a lasting impact on an organization. In order to find performance drivers that need changing, however, effectiveness must first be measured in terms of satisfied, loyal customers.

The influence of customers in selecting processes to improve was discussed in Chapter Two. In this chapter, customers are the starting point for discussing measures of effectiveness and the family of measures. The customer again becomes the focal point in Chapter Four, when we analyze the current situation (Step 2 of the Quality Improvement Model) beginning with the customer and chaining backward through the process that produces the output received by the customer.

MEASURING CUSTOMER SATISFACTION AND LOYALTY

Satisfied, loyal customers are measured using a variety of methods that address various categories of product and service quality. Focus Groups and Surveys (using interviews and Questionnaires) are the most commonly used methods (see

Chapter Eight: Collecting Data and Chapter Fourteen: Questionnaires). Other methods provide important but often less organized data, such as hotlines, suggestion and complaint letters, telemarketing call responses, and salesperson call reports. Teams need to consider the pros and cons of each tool relative to the situation in which it will be applied. The characteristics of the tool, the customer, and the type of data sought should all be considered. The most important recommendation is to collect only vital data.

Three broad categories of measures can be used to measure outputs and form the basis for customer satisfaction measurement: cost, delivery, and quality. Each should be measured in terms of the customer's definition. Cost measures relate to the price paid and the value perceived by the customer. Costs include all resources consumed, such as money, time, and materials. Delivery identifies both the timing and quantity measures for outputs, such as order cycle times and on-time delivery of ordered quantities. Quality measures fall into various dimensions of quality, or characteristics, which are summarized in Figure 3.1.

The eight dimensions of product quality were developed by David A. Garvin and the five service quality dimensions (denoted by an asterisk) were developed by Parasurman, Zeithaml, and Berry. The tangible service dimension refers to the appearance of the facility, equipment, and the personnel providing the service.

The listing in Figure 3.1 is presented not as a cookbook recipe, but rather as a guide to assist the reader in identifying appropriate categories to explore. Customer satisfaction measures generally are scaled responses for a specific category or for overall satisfaction. For example, the responses could rate satisfaction on a scale of 1 to 10, where 10 represents greatest satisfaction, or as a percent of customers with responses categorized as "satisfied."

Because failures are presumably the exceptional event, customer satisfaction can also be measured in terms of the absence of failures. For example, error-free invoices might be a measurable objective that leads to customer satisfaction.

Product	Service
Performance	Accessibility
Features	Reliability *
Reliability	Tangibles *
Durability	Responsiveness *
Conformance	Empathy *
Serviceability	Assurance *
Aesthetics	Credibility
Perceived Quality	Courtesy
	Competence

FIGURE 3.1 Dimensions of Quality

Organizations now understand that dissatisfied customers seldom complain or offer suggestions for improvement, but instead satisfy their needs and expectations elsewhere. As a result, organizations are developing creative ways to obtain feedback from all customers, particularly those who are dissatisfied. Customer loyalty is generally measured in terms of the customer's actions, such as repeat purchases, increasing activity, a wider array of product and services purchased, number of referrals, and percentage of total business.

FAMILY OF MEASURES

Customer satisfaction and loyalty measures are not the only measures needed for quality improvement. Quality teams need to think beyond end users, or external customers, and consider internal customers as well in a series of internal supplier–customer linkages that begin with external suppliers. Additionally, secondary customers, such as stockholders and regulatory government organizations, need to be considered. Although most work units do not "own" processes that have a measurable impact on shareholder value, for example, this should not be excluded from the team's consideration.

In Chapter Four, Step 2 of the Quality Improvement Model addresses the analysis of the current situation, which involves several key activities:

- Completion of a Process Analysis Worksheet (see Chapter Seven: Process Definition) that describes the relationship between the customer and the supplier in terms of specific performance measures.

- Development of Top-Down and Deployment Flowcharts (see Chapter Seven: Process Definition) which define the sequence and responsibility for performing activities in a process that produces an output.

- Preparation of a summary of key performance measures and attributes related to a process improvement opportunity. This involves a specific process with a known output. Data for the Table of Performance Indicators and Drivers (Figure 3.2) become available as the team follows steps in the improvement effort. The "As Is" column describes the current situation and the "Desired State" column identifies target measures after improvement. These conditions are discussed further in Chapter Four, Step 1: Select Improvement Opportunity.

These documents require an understanding of the family of measures. Because of limited time, quality teams should collect only the vital few measures necessary. Often the best measures are those related to failure, the exceptional event.

The discussion of the family of measures begins with the selection of an improvement opportunity, which is an important process in need of improvement. The first measures to address are outcomes, or the measures of organization effectiveness. The difficulty here is in identifying appropriate outcomes

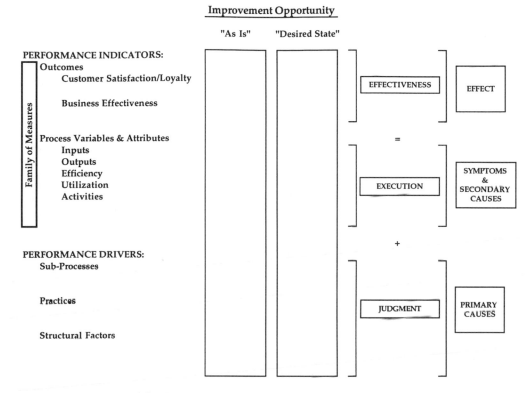

FIGURE 3.2 Table of Performance Indicators and Drivers

that are related to the outputs of the process selected. Only outcomes with an identifiable relationship should be included.

As discussed earlier, outcomes include both customer and business measures of effectiveness. Customer-related outcomes fit two primary categories: customer satisfaction and loyalty. Business outcomes are more broadly defined. They include such things as return on assets, profitability, market share, revenue growth, and measures that confirm achievement of specific organization goals and objectives.

Next, teams should focus on process variables and attributes. Here, the inputs to the process and the outputs from the process can be measured in terms of three primary categories: quality, cost, and delivery. For each customer–supplier relationship there is an output from the supplier, which becomes an input to the customer. Both the customer's needs and expectations and the supplier's specifications can be expressed in terms of these three categories. After negotiating mutually acceptable values, both parties can measure performance against some standard. Satisfactory relationships between supplier and customer depend on four common elements of these standards: they must be measurable, achievable, relevant to both parties, and controllable.

Finally, teams address other process-related measures, such as process efficiency and utilization, or productivity, and activity levels. The measures here are process specific and relate to the work unit performance, which is the execution portion of the effectiveness equation. Important measurements here include, for example, those that document process variation, process delays, non-value-added activities, and failures and waste.

Typical process measures include cost per unit, production rate per hour, sales per employee, labor productivity, cycle times (e.g., service response, design cycle, order lead times, and manufacturing cycle times), resource levels (e.g., staffing levels and training hours), on-time delivery, percentage defects, percentage scrap, number of suppliers, and activity levels at key points in the process.

Without changes in work practices, the process, or structural factors, improvements in outputs and outcomes are unlikely. In other words, the "as is system" is already defined by its processes and practices; improvements will occur only when the system is changed. The practices involved might be as simple as the technology applied or manual versus mechanical procedures. Structural factors might include organizational layers, span of control, and environmental considerations.

This brings us to the performance driver section of Figure 3.2. As the quality team follows the steps in the Quality Improvement Model, it will identify the root causes contributing to the unsatisfactory results. This provides information that enables completion of the performance driver section. The focus in this section is on qualitative characteristics of the subprocesses, practices, and structural factors that currently exist and need changing.

DATA TYPES

Throughout this book the word "data" refers to information that is both numerical and nominal (i.e., named or word descriptions). Definitions of the types of data are summarized in Table 3.1.

Appropriate data types for tools used to analyze and display data (Chapter Ten: Analyzing and Displaying Data) are listed in Table 3.2. Tools and techniques not listed generally use nominal data or the selection of data type is seldom a problem. Unless otherwise stated, tools with the data type "interval" display either continuous or discrete interval data. The data types shown are those most commonly used, although there may be exceptions.

DATA VALIDITY, RELIABILITY, AND BIAS

Because the data collected and analyzed in quality improvement are used to make decisions, it is important that they be valid and reliable and contain no biases. Validity indicates that what was measured is what was intended to be measured. It implies that the measures are both relevant and correct. Asking for information in observable and measurable terms improves data validity. Reliabil-

TABLE 3.1   Data Types

Data Type	Definition
Yes/No	Data with only two states; usually expressed in words, but often coded as 0 or 1 for computer analysis
Nominal	Data expressed in words; no interval or order is implied; numbers can be assigned to named categories
Ordinal	Sequential data expressed in words or numbers; intervals between data elements not uniform
Discrete (Interval)	Numerical data expressed as whole numbers separated by uniform intervals, without fractions or decimals; often obtained by counting; also referred to as attribute data
Continuous (Interval)	Numerical data expressed as fractions or decimals; obtained by comparison with a reference scale; also referred to as variable data

TABLE 3.2   Data Types for Analytical Tools

Tool	Data Type
Bar/Column Chart	Base line: Nominal, ordinal, or yes/no Scale line: Discrete interval
Pie Chart	Nominal labels Slices: Continuous interval percentages
Run Chart	Base line: Interval or ordinal Scale line: Interval
Histogram	Base line: Interval Scale line: Discrete interval
Pareto Chart	Base line: Nominal or ordinal Scale line: Discrete interval
Scatter Diagram	Base line: Interval Scale line: Interval
Control Charts	Base line: Ordinal or discrete interval Scale line: Continuous interval (Variable Charts) and discrete interval (Attribute Charts)
Process Capability	Base line: Continuous interval Scale line: Discrete interval

ity indicates consistency of measurement under stable conditions. That is, results can be repeated, given that everything else is the same.

Data can also suffer from biases during collection or analysis. If the team did not collect the data used, it needs to ensure that the data are not biased. Several key questions need to be answered:

- Who collected it?

- How was it collected?

- When was it collected?

- What assumptions were used?

In addition, the team needs to avoid personal and group bias in interpretation. This bias often takes the form of:

- Very quick decisions without discussion

- Filtering of data and ideas from outside the team

- Rationalizations using illogical explanations

Finally, the team needs to avoid "analysis paralysis," which is the result of too much, too little, and/or inconsistent data. The test of materiality should be applied to either discontinue the analysis if the data are not needed to make a decision or collect additional data if needed.

CHAPTER FOUR

QUALITY IMPROVEMENT MODEL

The Quality Improvement Model provides a common methodology and a common language for organizations to use in improving the quality of their products, services, work processes, and practices. Because quality teams seldom address isolated problems that are not part of a work process, quality improvement is primarily process improvement. Although the model is primarily intended for this purpose, it is easily adapted to solving stand-alone problems because it follows the Plan-Do-Study-Act problem-solving cycle. Teams that use the model for problem solving should eliminate steps related to work processes.

The background information in Chapters Two and Three helps to ensure that teams:

- Understand the voice of the customer

- Know the organization's goals and objectives

- Are familiar with the family of performance measures

- Understand that performance drivers must be changed to improve outcomes

This provides the foundation for applying the Quality Improvement Model. After reading this chapter, the reader should be able to understand and apply the process in the work environment. Each step in the process uses tools described in Chapters Five through Twelve. Additionally, this chapter includes references to the special tools covered in Chapters Thirteen and Fourteen.

Each step in the model includes suggested tools and techniques. Note that these are suggestions and not prescriptions. The reader should use judgment in selecting the appropriate tools that fit a particular situation. A matrix of process steps and tool categories is presented as a general guideline at the end of this chapter. This book explains commonly used tools and sets general boundaries for their use, but intelligent, empowered readers must use them wisely!

THE EIGHT-STEP MODEL

The Quality Improvement Model is an eight-step process that is performed in four phases (see Figure 4.1). The effectiveness of the improvement activity again depends on judgment in the Planning phase and on execution in the Do-Study-Act phases. Within each step, a logical flow of planning, analysis, study, and action occurs, which repeats the cycle of the overall model. The judgment in Planning is dependent on a customer focus and on using data to make decisions. It works best in an environment where employees are involved and empowered to act and where team efforts receive recognition.

In other words, it works best where the organization culture has shifted from the traditional to one that embraces quality management concepts. The model shares common elements with individual problem solving, but its main difference is its use by empowered, cross-functional teams focused on improving processes.

The model is not intended to be a recipe for success. It is a guide, or a logical process, that should be modified to fit the situation. Teams need this flexibility, particularly when barriers are encountered and the team needs to return to an earlier step.

Detailed activities, outputs, and results of the Quality Improvement Model are summarized at the end of each step.

Phase	Step
Plan	1. Select Improvement Opportunity
	2. Analyze Current Situation
	3. Identify Root Causes
	4. Select and Plan Solution
Do	5. Implement Pilot Solution
Study	6. Monitor Results and Evaluate Solution
Act	7. Standardize
	8. Recycle

FIGURE 4.1   Eight-Step Quality Improvement Model

STEP 1: SELECT IMPROVEMENT OPPORTUNITY

The purpose of this step is to define and select an important project and to obtain management's support for that project. If we look back at the ideas in Chapter Two, we see that the organization should begin its quality improvement by first seeking customer knowledge and then converting this knowledge into specific goals and objectives, which are then deployed throughout the organization. This logical approach is part of the assessment and the goal alignment of step zero. If the organization has not done this important work, then portions of step zero must be performed here.

For most organizations, quality improvement begins with a Steering Committee, a senior management team, a work unit, or an existing cross-functional team. The team starts by assembling currently available data that indicate that a process needs improvement. The data may be customer complaints, excess scrap, unexplained process variation, long delays caused by non-value steps in a process, or a competitive service that was not anticipated. The data are collected using a variety of tools, including informal and formal interviews, and Questionnaires, Focus Groups, and other reporting mechanisms.

The organization may have previously conducted both external and internal assessments to identify important processes. These assessments not only identify important processes that need improvement but also identify the organization's strengths (important activities, effectively performed), areas not needing change (low importance, low effectiveness), and areas possibly needing a reallocation of resources (low importance, effectively performed). Once the improvement opportunity is defined, the improvement team is formed, or in the case of an existing team, the team may be redefined.

Generate List and Select Important Improvement Opportunity

The tasks involved in selecting a problem or process improvement opportunity follow a systematic approach:

- Identify general and specific problems or opportunities.

- Clarify each one to ensure that team members understand them.

- Combine these problems or opportunities into related groups and identify high-priority ones based on rational selection criteria.

- Select the highest priority or most critical one and write a statement defining the current "as is" situation and the "desired state" in measurable terms.

The Matrix Diagram technique, Process Identification for Work Units, presented in Chapter Two works well in identifying processes needing improvement.

A common series of tools will be repeatedly used throughout the Planning phase to generate ideas and reach consensus (Chapter Five: Idea Generation and

Chapter Six: Consensus). Other tools can also be used, depending on the skills of the team. Several tools stand out in helping to select improvement opportunities:

- Criteria Rating Forms (Chapter Six)

- Matrix Diagram (Chapter Ten)

- Prioritization Matrix (Chapter Ten)

The Criteria Rating Form is easy to apply and helps to test the pulse of the team using understandable criteria. It is often used in conjunction with Matrix Diagrams that help to relate the "what" and "how" relationships: "what" is required to satisfy customer needs and expectations, for example, with the "how," or processes that deliver the outputs necessary to satisfy customers. This is followed by an evaluation of the effectiveness of each process, as described in the Process Identification technique presented in Chapter Two.

The Prioritization Matrix is a more sophisticated tool that combines the concepts of the first two tools. It lacks, however, the important element of evaluating how well a process is performed, that is, its effectiveness in achieving the desired outcome.

Several criteria useful in selecting projects (e.g., criteria used with the Criteria Rating Form) include:

- Controllable (i.e., within team control)

- Measurable results identifiable

- Achievable

- Data available or easy to capture

- Resource availability and requirements

- Significant importance, high visibility

- Timely completion

- Probability of success

- Team motivation and involvement

- Senior management support and commitment

Some organizations start their quality journey by documenting all processes in an extended planning phase. Although documentation (i.e., with Flowcharts and standard operating procedures) is important for various vendor certification requirements, it is often not the first objective if you want an immediate return on your training investment. The methodology presented assumes that the organization seeks immediate benefits from quality training by addressing high-priority ineffective processes first. This approach delays documentation of key processes until Step 2, Analyze Current Situation.

Redefine Team

The team that selected the improvement opportunity may need to be modified for the project itself. If a Steering Committee or senior management team made the selection, the actual team needs to be convened to explore the opportunity. Even in organizations with mature quality efforts where the majority of opportunities are identified at the work unit level, the team may need to be redefined to match the project needs. Effective teams generally consist of five to seven members with appropriate subject matter expertise.

Process "owners" should either be team members or act in supportive roles to help the teams succeed. Owners are individuals who have direct responsibility for achieving desired outcomes and for key activities in a process. Communication with external process owners is critical to success, primarily to avoid their resistance to change when solutions are recommended.

There are several key team roles that need to be filled: team leader, scribe, timekeeper, facilitator or process guide, and note-taker. Team roles are often rotated to recognize the importance of cross-functional stakeholders. This rotation aides consensus and avoids resistance later in the improvement process. Practice in a variety of roles also enhances team member skills.

Write Problem or Opportunity Statement

Writing a problem or opportunity statement helps to focus the team on the improvement opportunity. It is composed of two parts. The first part describes the "as is" or current condition, and the second part describes the "desired state," which reflects achievement of the improvement opportunity.

The "as is" portion of the statement defines the current conditions that suggest that a problem exists, a process needs improvement, or a practice needs to be changed. The "desired state" is the desirable outcome that might be achievable if the problem were solved, the process improved, or the practice changed.

The statement is a starting point that should be reviewed periodically as new information and data are uncovered. It should be a clearly defined, objective statement without biases, expressed in measurable, observable terms. The statement should not be too broad, nor should it suggest either causes or potential solutions. Furthermore, the statement should pass several key tests to ensure that the team's efforts will be effective:

- Is the problem or opportunity one that requires a team effort?

- Is it one that is worth solving or improving and is it critical to the organization?

- Are team members either stakeholders or owners of the process or practices affected? In other words, is it within their control or influence?

- Do the team members, individually or as a group, understand the problem or opportunity; are they subject matter experts?

- Does the team have sufficient data, or the potential for collecting data, to analyze the problem or opportunity?

Summarize Project/Define Road Map

The project summary and road map include the problem or opportunity statement, project scope, roles and responsibilities, meeting schedules, management status review dates, other milestone dates, and deliverables (i.e., physical outputs such as documents or products). The Is/Is Not Stratification technique (Chapter Six: Consensus) is helpful here in defining what is or is not included in the project scope. Several planning tools (Chapter Eleven: Planning Tools) are particularly useful here: Gantt Charts, Storyboards, and Tree Diagrams. Planning that requires more detailed tools usually indicates large projects with a broad scope.

The project summary can also include the methodology and rationale for selecting the project and barriers or issues that may be encountered, as well as risks and countermeasures. The concepts of the Process Decision Program Chart (Chapter Eleven) help in addressing these potential "what if" situations.

At this point, the project scope is expressed in general terms since the analysis of the current situation has not begun. The project scope should be one that can be easily completed in four to six months. Although some teams will start with a clearer picture of the situation (i.e., if the process has already been documented in an earlier project), most teams start without this documentation.

Two meeting management tools, Agendas/Meeting Minutes and Plus/Delta (Chapter Twelve), can be used to guide the team in conducting meetings and/

Step 1	Activity	Outputs and Results
Select Improvement Opportunity	Generate List and Select Improvement Opportunity	
	List and Select Important Opportunity Based on Criteria	Prioritized Project List
	Redefine Team	
	Write Problem/Opportunity Statement	Problem/Opportunity Statement
	Summarize Project/Define Road Map	Project Summary
	Management Review	Management Concurrence

FIGURE 4.2   Select Improvement Opportunity

or documenting its activities. Storyboards are particularly helpful in communicating team actions during management presentations.

Management Review

The purpose of the management review at this point is to ensure that the sponsor and/or customer of the project validates the thinking of the team in addressing the improvement opportunity. Tools used here, and in subsequent management reviews, include those applied in the respective step being reviewed. Graphic tools are preferred because they better illustrate relationships and key findings.

STEP 2: ANALYZE CURRENT SITUATION

The purpose of this step is to define the process to be improved in terms of its current configuration (e.g., using Flowcharts and operating policies or procedures), the participants in the process (i.e., customers and suppliers), and the performance measures that define customer expectations and supplier specifications. The team will use existing and additional data to narrow the focus of the study, to document baseline performance, and to identify gaps relative to expected performance.

Activities in this step often depend on the actions in step zero and other organizational assessments similar to the identification of performance drivers discussed in Chapter Three. The approach there was to begin with customer knowledge and work toward identifying performance drivers and the vital few measures of performance: desirable outcomes and process measures.

If key performance drivers are known, then this step is easier. Often, however, this is not the case for most work units. This step then begins with existing documentation or perception of an undesirable outcome or failure in the process. The team starts with this known effect and begins its journey toward finding the primary or root causes: the performance drivers.

As in other steps in the Planning phase, idea generation and consensus tools are used extensively. If the team is repeating a project previously addressed in continuous improvement, then Step 2 becomes a validation of previous documentation.

Define Process to Be Improved

Defining the process has several distinct steps that begin with a Process Analysis Worksheet (Chapter Seven: Process Definition) and conclude with a detailed Deployment Flowchart (Chapter Seven: Process Definition). Deployment Flowcharts not only show the steps in the process, but also identify the individuals or groups performing each activity (in separate columns or rows). These Flowcharts help to identify when and where delays, process variation, and failures occur, which highlights quality improvement opportunities and areas needing additional data.

The Process Analysis Worksheet includes the following items:

- Process Output
- Customer/Supplier Relationships
- Customer Needs and Expectations
- Performance Indicators
- Supplier Specifications
- Top-Down Flowchart
- Deployment Flowchart
- Table of Performance Indicators and Drivers

First, the outputs of the process to be improved need to be defined. The output is generally defined in terms of a noun-verb format that is not too broad or too vague (e.g., report completed) to be useful. The output should also not be defined so narrowly that it is insignificant, such as one that defines output from one activity in the process. Outputs emphasize completed work that is delivered to the customer in the form of a product or service. Outputs are not outcomes, such as satisfied customers, which may be the result of one or more processes.

Second, the customer needs to be identified. The primary customer is usually the next person or organization in the process that receives the process output. This customer may be either internal or external. Additional secondary customers (i.e., generally the next customer in line after the primary customer or an interested third party) are then listed. Individuals who approve or monitor outputs, such as department managers who approve departmental reports, are generally not the primary customer, however.

Each customer/supplier relationship must be explored because there may be conflicting needs and expectations among the customers. For example, the output may be used differently by different customers, which may result in a variety of favorable and unfavorable outcomes from similar outputs. Where appropriate, teams should think in terms of the external end user as the customer. Often, this customer is not accessible and a surrogate customer needs to be substituted to speak for other customers. Information from these surrogate customers can be obtained using Focus Groups.

Understanding how the customer defines the output and the resulting outcomes is the next task. If the team does not have this data, it must consult the customer and establish or verify these requirements: it must define the wants, needs, and expectations of the customer. (Again, the reader should look beyond "requirements" and gain knowledge of customer needs and expectations.)

These customer needs and expectations should be defined in specific, measurable terms and ranked in order of importance to the customer. The team must clarify and negotiate uncertain or conflicting requirements. The customer/supplier relationship must allow for this communication and negotiation of requirements if it is to be effective. Conflicts between output requirements and outcomes also must be resolved. Finally, needs and expectations should be expressed in terms of the three categories of measures discussed in Chapter Three: quality, cost, and delivery. The Table of Performance Indicators and Drivers presented in Chapter Three helps to summarize the vital few performance indicators that define outcomes and appropriate process variables and attributes.

Next, the team needs to translate customer needs and expectations into specific, measurable supplier specifications that define the process. Each requirement should have a corresponding measurable specification that is realistic and achievable. The need for cross-functional representation on the team becomes critical here. Where appropriate, teams then address inputs to the process and define input specifications, which are their customer requirements for their suppliers. The customer–supplier relationship chains backward using similar concepts and data requirements.

Written procedures that define the process (such as standard operating policies or procedures) can be helpful at this point but are not required for the next step: development of Top-Down and Deployment Flowcharts. To minimize the effort required, it is recommended that standard operating procedures be developed on a process, rather than functional, basis and that only procedures for key processes be done initially.

Collect Baseline Data and Identify Performance Gaps

Next, baseline data need to be collected from available sources to help define current performance (Chapter Eight: Collecting Data and Chapter Fourteen: Questionnaires). Data are collected for two purposes: to understand the process and possible causes of failures and variation and to establish the baseline of performance. Use both the Process Analysis Worksheet and then the Table of Performance Indicators and Drivers to summarize and display these data.

After the baseline data are collected, performance gaps are identified and the team is ready to continue with analyzing the root causes in Step 3. Before leaving this step, however, two additional activities need to be performed.

Validate Problem or Opportunity Statement and Management Review

This becomes the first of several possible areas where the team can validate the relevancy of the problem or opportunity statement based on new information. The team should now have a good idea of any revisions to the statement and the project scope.

Again, the Is/Is Not Stratification technique is helpful in defining the scope in more specific terms and keeping the project focused on important activities that are not effectively performed. Large projects may have to be broken into several pieces with the team separating into subteams or the scope may have to be narrowed for the team to be effective. Here too, the project road map should be modified to reflect these changes. The project scope should be one that can be easily completed in four to six months.

This is another milestone point to advise management of team progress.

Step 2	Activity	Outputs and Results
Analyze Current Situation	Define Process to be Improved	
	Identify Process Output	
	Identify Customer/Supplier Relationships	
	Identify Customer Needs and Expectations	List of Needs and Expectations
	Define Performance Indicators	
	Define Supplier Specifications	List of Supplier Specifications
	Flowchart the Process	Current Process Flowchart
	Collect Baseline Data	Baseline Performance Data ("As Is")
	Identify Performance Gaps	List of Gaps
	Validate Problem/Opportunity Statement	Revised Problem/ Opportunity Statement
	Management Review	Management Concurrence

FIGURE 4.3   Analyze Current Situation

STEP 3: IDENTIFY ROOT CAUSES

The purpose of this step is to identify potential root causes of poor performance, rather than secondary causes and symptoms, and to ultimately verify these root causes. All performance improvement depends on changing these root causes, the performance drivers: the processes, practices, and structural factors.

Activities in Step 3 are frequently iterative, beginning with the identification of potential causes, followed by data collection and analysis, more data collection, and finally verification of root causes. A well-designed data collection plan based on effective use of cause-and-effect tools can reduce the number of these data collection iterations. Here too, team members with subject matter expertise add significant value to the team process.

Analyze Cause-and-Effect Relationships; Identify Potential Root Causes

Identification of the root causes of undesirable outcomes is perhaps the most important activity in process improvement. It gets to the heart of what needs to be fixed or changed. This step in the model starts with the use of tools to analyze cause and effect and is followed by tools to collect and analyze data.

Three basic tools are used to explore the potential causes of performance gaps identified in Step 2: Cause-and-Effect Diagrams, Five Whys, and Interrelationship Digraphs (Chapter Nine: Analyzing Cause and Effect). The Matrix Diagram can also be applied here.

Benchmarking, which is an externally focused search for best practices, might be initiated here depending on the situation (see Chapter Thirteen: Benchmarking). If other organizations perform functionally equivalent activities, they are potential Benchmarking partners. The external view from Benchmarking often provides the spark for creative imitation, and it can validate root causes and proposals for change. Benchmarking can also be initiated in Step 4: Select and Plan Solution. Although Benchmarking could be used earlier in the model, it is best to analyze your own situation and establish baseline data first.

Collect Data to Verify Cause-and-Effect Relationship and Root Causes

Once the team has identified potential causes, it needs to establish a data collection plan that defines what data are needed, where and how they should be collected, how long they should be collected, and by whom. Areas of potential bias should be addressed before the plan is implemented. The need to collect statistically significant, unbiased data is a common requirement in quality improvement.

Data collection plans often start with Checksheets to collect raw data, followed by analysis using Pareto Charts, Histograms, Scatter Diagrams, and various graphs (see Chapter Eight: Collecting Data and Chapter Ten: Analyzing and Displaying Data). The primary objective here is to identify the strength of relationships between variables, discern patterns within data, and understand the characteristics of data.

Only the vital few performance indicators (outcomes and process variables and attributes) need to be addressed. Flowcharts developed earlier help to identify where to start looking for these vital few measures. Data that document failure points, waste, steps with significant variations, illogical work and information flows, and steps that add no value are the first to be investigated. Data sought here fall into three broad categories, or the three "V's" of process improvement:

- Reduction in **variation** that first addresses special causes and then common causes. Sources of variation that have a significant effect on process performance occur in the early stages of the process, during important steps in the process, in steps just before major points of failure or rework, or where manual practices and interfaces exist.

- Elimination of **non-value** activities, such as delays in decisions, unnecessary reviews of documents and reports, and in operations that are not line balanced (i.e., where bottlenecks exist).

- Increases in **velocity** (e.g., reduction in cycle time) by replacing sequential flows with parallel flows with concurrent reviews, by eliminating non-value steps and by simplifying processes.

Before verifying root causes, the data need to be validated for accuracy (i.e., complete and unbiased) and purpose. Insufficient data to verify the cause-and-effect relationship will require the collection of additional data, the iterative process discussed earlier. Finally, the data can be analyzed to verify the link between the performance indicators and the drivers to complete the search for root causes.

At the end of Step 3, the team identifies root causes and may choose to begin using Run Charts, Control Charts, and Histograms (Chapter Ten: Analyzing and Displaying Data) to establish baseline control data to monitor process improvement activities. The use of these tools will carry through the implementation phases of the model (i.e., Do-Study-Act) and may become permanent depending on future needs.

Statement Validation and Management Review

Again, the problem or opportunity statement needs to be validated again and management should be advised of team progress.

Step 3	Activity	Outputs and Results
Identify Root Causes	Analyze Cause and Effect Relationships	
	Identify Potential Root Causes	
	Collect Data	
	Verify Cause and Effect and Root Causes	Reasons for Gaps: Graphic Displays Linking Causes and Effects
	Validate Problem/Opportunity Statement	Revised Problem/ Opportunity Statement and Road Map
	Management Review	Management Concurrence

FIGURE 4.4   Identify Root Causes

STEP 4: SELECT AND PLAN SOLUTION

The overall purpose of this step is to identify and select the solution to solve the problem or improve the process based on the analysis in Step 3.

This step is performed in two parts. First, the potential solutions for process improvement are analyzed and selected based on various criteria, and second, an implementation plan is developed. The objective here is to select the best possible solution based on the data and facts available (i.e., select the solution with the greatest impact on the root causes). In addition, an implementation plan is developed that is sufficiently detailed to be actionable and that considers countermeasures for reasonable barriers to successful implementation.

As in other planning steps, the idea generation and consensus tools are used extensively to help the team identify and select the appropriate solution.

Generate List and Select Best Solution

Before selecting a solution, the team must be sure that the proposed changes are supported by the data collected and the analysis performed on the data in Step 3. Once this validation is complete, the team selects solutions using the same tools and techniques (particularly Criteria Rating Forms, as well as Matrix Diagrams and Prioritization Matrices) used to select the improvement opportunity.

A key criterion in process improvement is the robustness of the revised process, that is, how immune it is to uncontrollable factors. Techniques such as those used in Program Decision Process Charts to identify "what if" contingencies become a critical test of the robustness of the planned solution. These tests of robustness are used throughout this step, first to test the improved process and then to test the implementation plan. The contingency plan is based on anticipated barriers that can hinder success and factors that can enable or assist the implementation.

Define Revised Process

Once the best solution has been selected, the impact on the process needs to be documented to reflect changes in the process defined in Step 2 above. The key tasks here include:

- Identify Expected Outcomes

- Revise Output and Supplier Specifications

- Identify Target Process Values

- Modify Deployment Flowchart

The solutions involve changes in performance drivers, the root causes, rather than just targets aimed at the secondary causes and symptoms. The Table of Performance Indicators and Drivers can now be completed with target values for outcomes, outputs, and process data. These targets become useful in monitoring progress as the performance drivers change. Structural factors that cannot be changed should be acknowledged to ensure that the organization understands and accepts the structural barrier.

Modifications in outputs lead to revised supplier specifications that define the new standards of performance necessary to satisfy customer expectations and achieve desired outcomes. Various process values are targeted to ensure that the process can be monitored at different levels. Next, the team modifies the original Deployment Flowchart to reflect the planned solution.

Keeping customers advised or testing reactions to change before implementation helps to eliminate resistance. Frequent communication with process stakeholders also helps to minimize the resistance to change.

Develop Implementation Plan

The detailed implementation and contingency plan is a major milestone in the improvement process. This plan should provide periodic monitoring of performance indicators (i.e., both outcomes and process measures) to ensure that expected changes occur. Elements of the plan should describe:

- Sequence and Timing
- Resources and Controls
- Responsibility
- Pilot Activities
- Contingency Actions

It is important that the plan identify expected measurable results as well as clarify responsibilities, milestone reviews, and contingencies to respond to unexpected events. In summary, the plan should answer what, how, who, when, and where.

The most commonly used planning tools are Tree Diagrams and Gantt Charts (Chapter Eleven: Planning Tools). Storyboards are used more frequently here for documentation and communication rather than for planning purposes. Large, complex projects or projects with great uncertainty often require more sophisticated tools such as the Activity Network Diagram and the Process Decision Program Chart, respectively.

Management Review

This management review is perhaps the most important one in the improvement process. Up to this point, no decisions to change the organization have been made. Here, the team presents its findings and, more importantly, makes recommendations for change. The primary objective is to gain management's concurrence and approval to continue. If the team has kept management advised of its progress, then the team should not have difficulty in gaining management's concurrence.

This review provides the team with the ultimate validation of management's commitment to the quality improvement process and the empowerment of employees. If management passes the test, the process continues and team members remain committed to continuous improvement. Unrealistic barriers erected by management at this point make the internalization of continuous quality improvement difficult at best.

It is the team's responsibility to keep management informed of its progress, but it is management's responsibility to ensure that interim reviews are conducted and to provide clear, concise, and candid feedback. This feedback helps to prevent unrealistic expectations that can result in disappointment when recommendations are rejected.

Finally, the implementation plan should provide a road map of the remaining steps in the improvement process, including any training that will have to be developed for implementation after the pilot activity.

Step 4	Activity	Outputs and Results
Select and Plan Solution	Generate List and Select Potential Solutions	
	List Solutions and Select Best One Based on Criteria	Prioritized Solution List
	Define Revised Process	
	• Revise Process Output	
	• Identify Expected Outcomes	Performance Targets ("Desired State")
	• Revise Supplier Specifications	Revised Specifications
	• Modify Flowcharts	Modified Flowcharts
	Develop Implementation Plan	Detail Implementation/ Contingency Plan
	• Identify Sequence/Timing	
	• Define Resources/Controls	
	• Define Responsibility	
	• Identify Pilot Activities	
	• Identify Contingency Actions	
	Management Review	Management Concurrence

FIGURE 4.5   Select and Plan Solution

STEP 5: IMPLEMENT PILOT SOLUTION

Conduct Pilot

The purpose of a pilot implementation is to test the solution on a small scale in order to ensure that the revised process is capable of producing an output with the desired outcomes. The pilot activity verifies the effectiveness of the solution

Cascade Beyond Pilot Activity Using Appropriate Training

A key activity included in this step is the development of training materials to enable the standardization. This training will build upon the knowledge and experience gained by the pilot organization and often requires preparation of formal training materials. Training organizations, which are generally not part of the improvement team, are critical to the success of the standardization. They should be brought into the improvement activity as soon as feasible (e.g., no later than during implementation planning, Step 4).

Training developed here will be based on new policies and procedures or on changes to existing ones affected by the process improvement. Good ideas need to be communicated well, and training is a key element of success in standardizing process improvement.

Monitor Results and Evaluate Solutions

The monitoring and evaluation of results need to continue to ensure that the desired results observed in the pilot are achieved by all organizations affected. The benefit of selecting the vital few measures to track performance is clearly illustrated here: data collection should be a means to an end rather than an end itself. Tools used in Step 5 are used here as well, but on a broader scale.

Document Entire Quality Improvement Journey

Next, the team needs to complete the road map that documents and communicates its journey. Documentation can be in various forms, including written reports, data files, abstracts in project tracking systems, videotapes, and Storyboards. Documentation should meet the minimum requirement of enabling subsequent teams to learn from their experience and continuously improve the same process.

Finally, the team should determine long-term monitoring requirements and future actions. The team should meet again when unsatisfactory changes in performance measures or outside trigger events (e.g., new technology, products, or services) occur or a specified time has elapsed.

Management Review

Before leaving this step, the team needs to conduct a final management review to acknowledge closure, to adjourn the team activities, and to provide management with an opportunity to recognize and reward the team. Once the organization has fully implemented the changes, it will be ready to tackle the next most important issue.

Step 7	Activity	Outputs and Results
Standardize	Cascade Beyond Pilot Activity	
	Develop Appropriate Training Materials	Training Curriculum
	Monitor Results and Evaluate Solutions	Progress Report
	Document Entire Quality Improvement Journey	Documented Summary of Project
	Management Review	Management Concurrence

FIGURE 4.8 Standardize

STEP 8: RECYCLE

The purpose of this final step is to ensure that the organization becomes focused on continuous improvement. Step 8 becomes Step 1 in the next cycle of quality improvement.

Identify New Improvement Opportunity

Team members should begin to address the next most critical problem or opportunity in their work unit. Team members often participate in multiple projects in various phases of quality improvement, addressing activities both within their work unit and activities that cross multiple functions.

As quality management matures in the organization, the elimination of departmental barriers and the focus on processes will likely result in work units being organized around specific key processes that include multiple functions. Continuous improvement can then be performed by semi-permanent teams that have a stake in these key processes that define their work unit.

Step 8	Activity	Outputs and Results
Recycle	Identify New Improvement Opportunity	Revised Opportunity List

FIGURE 4.9 Recycle

QUALITY IMPROVEMENT TOOLS AND TECHNIQUES MATRIX

Figure 4.10 is presented as general guide for selecting tools and techniques to apply in the eight steps of the Quality Improvement Model. Tools and techniques for each step in the model have been suggested in this chapter, but specific prescriptions were avoided. It is up to intelligent, empowered readers to make the right choice for their situation!

Several authors, however, feel compelled to be specific. One offers a matrix of 99 tools that gives specific recommendations for each of twelve problem-solving phases. Of the 99 tools, 24 are indicated to be useful in reaching agreement. The dilemma becomes which tool to use to agree on the right tool for reaching agreement.

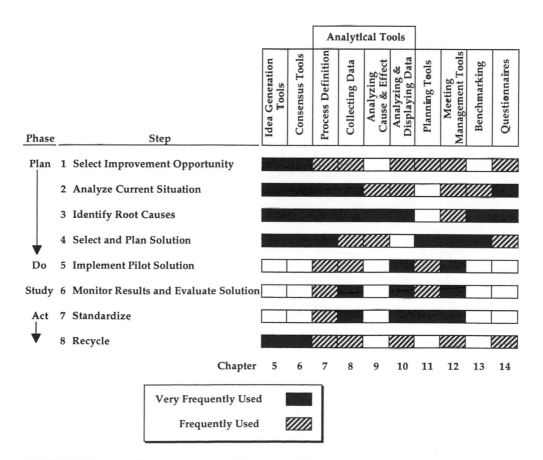

FIGURE 4.10   Quality Improvement Tools and Techniques Matrix

Chapter Five

IDEA GENERATION

The planning steps of quality improvement rely on a succession of idea generation activities followed by convergent activities culminating in consensus at each decision point. The primary tools used for generating ideas are covered in this chapter.

The expansion activity uses intuitive, right brain thinking, where ideas are somewhat random and jumbled. The purpose of this phase is to generate many ideas and encourage the creativity and diversity of the team. During idea generating, communication centers on initiating ideas and piggybacking ideas. This is followed by logical, left brain thinking, where ideas are clarified, sorted, and combined into organized categories. The team is then ready to begin the process of reaching consensus.

A typical process for defining a problem or improvement opportunity during Step 1 of the Quality Improvement Model, Select Improvement Opportunity, might take the form of the following sequence of activities:

- Brainstorm a list of potential problem areas or improvement opportunities

- Clarify each item on the list without discussing the merits of the idea

- Reduce the list to a manageable number (say ten items) using List Reduction, which is where the evaluation and discussion of ideas occur

    o Combine similar items, cross out duplicates

    o Set aside alternatives (i.e., decide to address less important ideas later) based on informal selection criteria and Weighted Voting

- Continue reduction using more restrictive and formal criteria

- Develop "as is" statement for top two or three alternatives

- Use Criteria Rating Form to guide selection of one problem or opportunity
- Develop "desired state" portion of problem statement

The criteria used to select a problem or opportunity (next to last activity above) generally include:

- Controllable (i.e., within team control)
- Measurable results identifiable
- Achievable
- Data available or easy to capture
- Resource availability and requirements
- Significant importance, high visibility
- Timely completion
- Probability of success
- Team motivation and involvement
- Senior management support and commitment

The idea generation tools include:

- Brainstorming
- Brainwriting
- Crawford Slip Method

BRAINSTORMING

Description

Brainstorming is an idea generation technique that stimulates creativity, encourages team participation, and minimizes premature critiquing and evaluation of ideas. Participants are encouraged to think opposites, add wild ideas, and think "out-of-the-box" since all ideas are recorded.

Key Points

- Encourages everyone to participate and to be creative
- Allows participants to piggyback off each other's ideas
- Generates a large number of ideas
- Two primary forms: free-wheeling (i.e., no specific order of participation) and round robin (i.e., participants contribute in some order)

- Free-wheeling is more spontaneous and allows easier piggybacking, but can be confusing and one individual can dominate

- Round robin is more orderly and allows uniform participation, but is less spontaneous and limits piggybacking

Typical Applications

- Generate list of potential problem areas

- Generate list of improvement projects

- Identify possible causes of a problem

- Generate list of potential solutions

Example

```
Perceived Problems/Improvement Opportunities
              Low return on sales
              Long check-out time
              Check-out process not efficient
              Not enough titles
              Store layout poor
              Insufficient display space
              Wrong movies in stock
              Too few science fiction movies
              Too few copies of new releases
              Declining profits
              Poor selection
              Too long to check out
              No drop box for after hours
```

FIGURE 5.1 Brainstorming: Video One Problems/Improvement Opportunities

Steps

1. State the topic of the brainstorming session; clearly write the topic at the top of a flip chart or board.

2. Explain the rules of Brainstorming to the participants:

- All ideas are recorded; evaluating or criticizing ideas is not allowed.

- Participants may piggyback (i.e., add to an existing idea to create another idea) and offer more than one idea at a time while free-wheeling.

- Participants must wait their turn to add new ideas or to piggyback while using round robin; they can pass if they have no new ideas.

3. Ask for questions regarding the rules and obtain agreement on the method (free-wheeling or round robin) to use.

4. Record all ideas on a board or flip chart in large enough text that everyone can see it. [If the team decides to use an Affinity Diagram to help reach consensus, they should use Brainstorming, but should select a second scribe to write ideas on 3 × 5 inch index cards (or Post-It™ notes) for later sorting. Teams might prefer to prepare the index cards after they have used List Reduction to eliminate duplicate ideas or alternatives that are set aside for later review.]

5. Encourage participants to state ideas in a succinct form to aid the scribe.

BRAINWRITING

Description

Brainwriting is an idea generation technique that stimulates creativity and encourages team participation. It is very similar to Brainstorming, but differs in that it relies on written rather than verbal communication. As in Brainstorming, participants are encouraged to think opposites, add wild ideas, and think "out-of-the-box." Brainwriting usually involves idea generation for multiple categories of a single topic.

Brainwriting has many variations, some of which do not rely on piggybacking ideas, such as the Crawford Slip Method. It can be a more focused approach than Brainstorming, resulting in more fully developed ideas that cover multiple categories of a topic.

Key Points

- Encourages everyone to participate and be creative
- Allows participants to piggyback off each other's ideas
- Generates a large number of ideas
- Allows for the development of ideas covering multiple topics
- Two primary forms: index cards and gallery method
- Both forms allow different topics for each card or flip chart

- Index card method: ideas are written on cards and circulated for the addition of related ideas

- Gallery method: ideas are written on multiple flip charts and participants move around the room to add related ideas

Typical Applications

- Generate list of potential problem areas

- Generate list of improvement projects

- Identify possible causes of a problem

- Generate list of potential solutions

Example

Problem: Title Selection	Problem: Customer Convenience
Not enough titles	Long check-out time
Wrong movies in stock	Check-out process not efficient
Too few science fiction movies	Not enough titles
Too few copies of new releases	Store layout poor
Poor selection	Poor overall selection
	Too long to check out
	No drop box for after hours

FIGURE 5.2   Brainwriting: Video One Title and Customer Convenience Problems

Steps

1. State the topic of the Brainwriting session; clearly write the topic and specific category at the top of the flip chart or each index card.

2. Explain the rules of Brainwriting to the participants:

   - All ideas are recorded; evaluating or criticizing ideas is not allowed.

   - Participants are encouraged to add new ideas or piggyback (i.e., add to an existing idea to create another idea) circulating cards or while viewing each flip chart.

   - All recorded ideas should be legible, with text on flip charts large enough for everyone to see.

   - Participants may write more than one idea at a time.

3. Ask for questions regarding the rules.

4. Encourage participants to write ideas in a succinct form.

5. Transcribe ideas from index cards to flip charts for the next step: reaching consensus.

CRAWFORD SLIP METHOD

Description

The Crawford Slip Method is an idea generation technique that stimulates creativity while drawing on the knowledge and views of participants. It is similar to Brainwriting but differs in that it protects the anonymity of participants and does not use piggybacking ideas. Like Brainwriting, it can be a more focused approach than Brainstorming, resulting in more fully developed ideas that cover multiple categories of a topic.

Crawford Slip is often used when teams are in conflict, when the subject is controversial, or when large amounts of information will be processed. (Several computer programs use this concept to protect anonymity while allowing participants to piggyback ideas.) In its simplest form, it can be used to anonymously capture ideas on index cards that address one subject.

Key Points

- Gathers ideas quickly and anonymously without group interaction

- Generates a large number of ideas

- Allows for the development of ideas covering multiple categories of a topic

- Piggybacking of ideas not utilized (except in computerized versions)

- Leader prepares statements and questions for participant responses on each category of the topic

- Participants' responses reflect their knowledge of and views on the subject

Typical Applications

- Generate list of potential problem areas

- Generate list of improvement projects

- Identify possible causes of a problem

- Generate list of potential solutions

- Prepare in-depth documents addressing the above applications

Example

Problems/Improvement Opportunities

Problem #1: Title Selection

1A. What are the important steps in selecting movie titles for rent?

1B. Why are we failing to select titles sought by our customers?

1C. Do our customer surveys correctly reflect their preferences?

1D. If not, how would you change them?

Problem #2: Customer Convenience

2A. What convenience items are important to our customers?

2B. How effectively are we satisfying convenience expectations?

2C. Do our customers complain about convenience items?

2D. If not, how can we determine if they are not satisfied?

FIGURE 5.3   Crawford Slip: Video One Title and Customer Convenience Problems

Steps

1. Leader prepares statements and questions to focus participant responses. For each category of the topic, participants are asked to respond to several statements and questions. These questions and statements should be coded for easy tabulation of responses.

2. Explain the following rules to participants:

   - Everyone is encouraged to respond to all questions and statements.

   - Multiple responses to individual questions or statements are encouraged.

   - Use 3 × 5 inch (or larger) index cards as follows:

     o Write across the top, long edge of the card.

     o Code each response card by topic and question or statement number.

     o Write one sentence per card; use short, simple sentences.

     o Write explanations and clarification on the reverse side of the card.

     o Use wording understood by those who lack subject matter expertise.

     o Write out acronyms the first time they are used.

     o   All participants are encouraged to write until time is called.

     o   All responses should be legible.

3.   Participants are given specific time limits (usually 20 to 30 minutes) to complete responses; participants do not have to respond to all statements or questions.

4.   Ask for questions regarding the rules or process.

5.   Sorting and consolidation of responses is performed next. This difficult task is usually not performed by the participants because of the volume of data and the amount of effort required. Once the information is sorted and consolidated into main categories, an outline is developed. This outline is then used to transcribe the responses into a final document.

6.   The final document can then be used by the team in the next step: reaching consensus.

CHAPTER SIX

CONSENSUS

The planning steps of quality improvement rely on a succession of idea generation activities followed by convergent activities culminating in consensus at each
decision point. The primary tools used for reaching consensus are covered in this
chapter.

After generating ideas, teams use logical, left brain thinking, where ideas are
clarified, sorted, and combined into organized categories. As teams discuss
options, their communication is characterized by summarization, seeking or
providing information, confirming understanding, and agreeing or disagreeing
on responses. The objective is to select ideas for further consideration based on
rational criteria, either generally understood or formally defined, and to reach
consensus.

Consensus is reached when everyone:

- Has voiced their opinion

- Is willing to support the decision even though it may not be their first
  choice

- Will accept, live with, and not oppose the decision

Consensus does not necessarily mean that there is unanimous agreement and
that everyone will be satisfied. The tools used in reaching consensus are not
intended to make a decision, but rather are intended to take the temperature of
the team. If all team members can answer the following question affirmatively,
then consensus has been reached:

> "Can you live with, actively support, and not oppose this decision
> even though it may not be your first choice?"

The process of reaching consensus after the generation of many, often unrelated, ideas can be a challenging task. Consensus can only begin after all ideas

have been clarified so that everyone has an understanding of each one. Consensus often takes longer than decisions made by one person, but it provides a level of commitment that is lacking under most decision-making methods. Teams with good communication skills and active participation of all members are usually more successful in reaching consensus.

The process of sorting and combining ideas uses several tools to filter or reduce the list down to the important items. List Reduction and Affinity Diagrams help to group ideas, while Paired Comparisons, Weighted Voting, and Criteria Rating Forms help to rank alternatives in order of their relative importance.

The Criteria Rating Form is particularly useful in helping teams reach consensus. The advantage of this tool is that team members first agree on several criteria for evaluating alternatives; the criteria are often less controversial than the alternatives themselves. Once the criteria are in place, the alternatives can be more easily discussed on a logical basis rather than a potentially emotional, subjective basis without the criteria.

The Balance Sheet and Is/Is Not Stratification techniques help to clarify and define differences in alternatives while striving for consensus. The Is/Is Not Stratification technique helps teams not only define the problem or improvement opportunity, but also define the project scope and, later in the quality improvement effort, stratify data for analysis.

A typical process for defining a problem or improvement opportunity during Step 1 of the Quality Improvement Model, Select Improvement Opportunity, might take the form of the following sequence of activities:

- Brainstorm a list of potential problem areas or improvement opportunities

- Clarify each item on the list without discussing its merits

- Reduce the list to a manageable number (say ten items) using List Reduction, which is where the evaluation and discussion of ideas occurs

  o Combine similar items and cross out duplicates

  o Set alternatives aside (i.e., decide to address less important ideas later) based on informal selection criteria and Weighted Voting

- Continue list reduction using more restrictive and formal criteria

- Develop "as is" statement for top two or three alternatives

- Use Criteria Rating Form to guide selection of one problem or opportunity

- Develop "desired state" portion of problem statement

The consensus building tools are:

- Affinity Diagram

- Balance Sheet

- Criteria Rating Form

- Is/Is Not Stratification
- List Reduction
- Paired Comparisons
- Weighted Voting

AFFINITY DIAGRAM

Description

The Affinity Diagram is a management and planning tool that is used to translate large numbers of complex, apparently unrelated ideas, issues, or opinions into natural and meaningful groupings of visual data. Grouping related items helps to identify underlying relationships that tie groups together. It is both a creative and a logical process in which consensus is reached by visual rather than verbal means.

The origin of the Affinity Diagram can be traced to another data analysis method, the KJ Method, developed during the 1960s by Japanese anthropologist Jiro Kawakita. The technique helped in the analysis of massive amounts of data in order to discover new patterns of relationships in the information. This tool is similar to Storyboards (Chapter Eleven: Planning Tools), which start with header cards and generate ideas related to the headers, the reverse of the process used in the Affinity Diagram. Storyboards predate the KJ Method, however.

Key Points

- It is both a creative and a logical method
- Promotes the emergence of breakthrough thinking
- Requires team participation with varied perspectives and open-minded creativity
- It is particularly useful in organizing large numbers of complex, apparently unrelated ideas, issues, or opinions; it is useful where apparent chaos exists
- It is also useful in grouping verbal data
- Also referred to as the KJ Method

Typical Applications

- Identify and group potential problem areas or improvement projects
- Identify possible causes of a problem prior to using an Interrelationship Digraph or other cause-and-effect tool (Chapter Nine)

- Group complex, apparently unrelated items into useful groups
- Identify and group related elements (e.g., tasks, ideas, functions, and requirements) for analysis with Tree Diagram (Chapter Seven) and for planning with Storyboards (Chapter Eleven)

Example

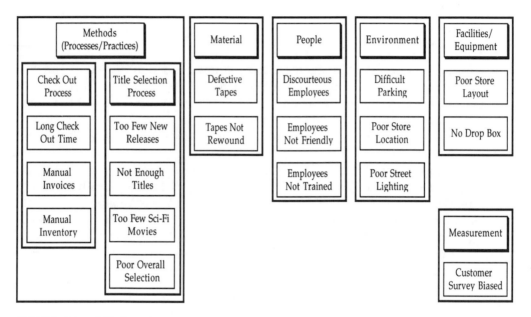

FIGURE 6.1   Affinity Diagram: Causes of Dissatisfied Video One Customers

Steps

1. Select a topic to be addressed. (For example, an "as is" problem or opportunity statement might be used to help identify and group potential causes that need to be explored.) State the issue in neutral terms, write it at the top of a flip chart or board, and underline it.

2. Explain the rules for Brainstorming. Encourage everyone to participate.

3. Use Brainstorming to generate ideas and record them on a board or flip chart *and* on 3 × 5 index cards or Post-It™ notes. The writing on the board or flip chart should be legible and large enough for everyone to see. Avoid one-word cards; short statements with a noun and a verb are preferable.

4. Randomly arrange index cards or Post-It™ notes on a large flat surface (wall or table). **Do not try to place the notes in a specific order.** Clarify all items without discussion or evaluation.

5. Sort cards into related, natural, or logical groupings **in silence**. This is done by participants based on their gut feelings, without verbal discussion. Cards can be grouped and regrouped by anyone at any time.

    Depending on the number of ideas, this activity may take an hour or more unless a time limit is established. However, it is critical that enough time be provided for everyone to contribute to the process. When inadequate time is provided, subsequent steps are more difficult for the team to complete. An alternative is to attach cards to a wall in a conference room where participants can rearrange the cards over a longer time period (e.g., 8 hours, 24 hours, or several days).

6. Discuss groupings after participants reach consensus on the final groupings (i.e., no new moves). Minor additional rearranging may occur at this time as the participants clarify their understanding of what has been accomplished.

7. Create header cards that describe the grouping. Headers can be generated from one of the items in the group or can be new statements. Total number of headers should be limited to five to ten. Avoid one-word headers. Separate large groupings into subgroups and create subheaders as necessary.

8. Draw a line around each grouping to indicate affinity with the header or subheader.

9. Compare results with original issue or problem.

10. Review final Affinity Diagram with others who did not participate; team should discuss and reach consensus before making modifications.

BALANCE SHEET

Description

A Balance Sheet is a tool to help teams identify the pros and cons of alternatives in their attempt to reach consensus. Again, the tool is used to promote discussion and test the team's position regarding the alternatives rather than make a decision.

This tool is generally used when the list of alternatives has been reduced to a manageable level. Balance Sheets are useful in identifying criteria for Criteria Rating Forms. The Balance Sheet is similar to the Plus/Delta (+/Δ) Evaluation technique used to evaluate meetings (Chapter Twelve: Meeting Management Tools) and the Force Field Analysis, which is a more analytical tool (Chapter Ten).

Key Points

- Uses Brainstorming to generate the pros and cons of alternatives

- Generates a large number of ideas

- Promotes discussions about divergent views and tests the team's feelings regarding alternatives

Typical Applications

- Prioritize, evaluate, and select problem areas or improvement opportunities

- Prioritize, evaluate, and select solutions

- Generally used after list of alternatives has been reduced using List Reduction

- Helps in developing criteria for use in Criteria Rating Forms

Example

Pros and Cons of Improvement Opportunities		
Project Alternatives	**+**	**−**
Check-out process	Customer convenience Important	Difficult to correct Lack resources for capital
Title selection process	Satisfies customers Important	Difficult to correct
Store layout	Customer convenience	Lack resources for capital Difficult to correct quickly
Computer controls	Better control	Lack resources for capital

FIGURE 6.2 Balance Sheet: Pros and Cons of Video One Improvement Opportunities

Steps

1. Use Balance Sheets when the number of alternatives has been reduced to a manageable level (i.e., less than ten).

2. Set up two wide columns on the right side of a flip chart or board; allow sufficient space to identify the alternatives in a column on the left side.

3. Place a "+" and a "−" as headers of the wide columns ("+" on the left and "−" on the right).

4. Follow free-wheeling Brainstorming rules. Encourage everyone to participate.

5. State the first alternative to be evaluated and write it in the narrow column at the left. Discuss one alternative at a time.

6. Record all ideas in large enough text that everyone can see.

7. Discuss results. Some ideas may be both pros and cons.

CRITERIA RATING FORM

Description

Criteria Rating Forms help teams reach consensus through the use of weighted criteria to rate alternatives. These criteria provide a logical basis for evaluating alternatives and are often less controversial than the alternatives themselves. The numerical results of Criteria Rating are not used to make decisions; they are used to quantify team member positions and preferences as the team tries to reach consensus.

Criteria Rating is a more formalized method of List Reduction where the criteria have clearly defined, numerical values. It is generally not the first tool used after idea generation; often, it is the last tool used before consensus is reached.

Criteria Rating Forms are similar to Prioritization Matrices (Chapter Ten: Analyzing and Displaying Data), but are significantly less difficult to use in reaching consensus. Prioritization Matrices are well suited to complex management decisions that require detailed analysis of many options. The analytical requirements of Prioritization Matrices influenced the decision to place them in Chapter Ten instead of this chapter.

Key Points

- Encourages everyone to participate
- Compares all alternatives using the same criteria
- Uses criteria that are often less controversial, and therefore easier to reach consensus on, than the alternatives
- Enables the weighting of important criteria
- Provides a logical basis for reaching consensus necessary for a decision

Typical Applications

- Prioritize, evaluate, and select problem areas or process improvement opportunities

- Prioritize, evaluate, and select solutions

- Generally used after list of alternatives has been reduced using List Reduction

Example

CRITERIA 1 Scale 5	WEIGHTING FACTOR	ALTERNATIVES			
		Survey Design & Analysis Process	Check Out Process	Title Selection & Ordering Process	Advertising Process
CONTROL Low High	1	5	4	4	3
IMPORTANCE Low High	3	2	5	4	3
DIFFICULTY High Low	1	4	3	4	4
RESOURCES High Low	1	4	4	4	3
PAYBACK Low High	2	3	4	3	2
TOTAL		25	34	30	23

FIGURE 6.3   Criteria Rating Form: Rating of Video One Improvement Opportunities

Steps

1. List the problem statements, process improvement opportunities, or solution alternatives as column heads across the top of the form.

2. Clarify all alternatives before selecting criteria. Use List Reduction or majority voting to limit alternatives to no more than ten before using Criteria Rating.

3. Discuss and select five or six criteria appropriate to the alternatives. Encourage participation in selecting criteria and in establishing criteria weighting. List criteria as row headings.

Sample criteria are:

- Problem areas or improvement opportunities:
  - o Control—Team control of solution
  - o Importance—Urgency of the problem
  - o Difficulty—Relative difficulty
  - o Time—Time available
  - o Resources—Resources required
  - o Payback—Return on investment of resources
- Solution alternatives:
  - o Control—Team control of solution
  - o Appropriateness—Satisfaction of requirements
  - o Acceptability—By organization
  - o Time—Time available
  - o Resources—Resources availability
  - o Payback—Return on investment of resources

4. Ensure that all team members understand the meaning of each criterion. Encourage everyone to participate.

5. Discuss and select the Criteria Rating format (i.e., rating scales or point scoring). Rating scales of 1 to 5 (where 5 is most important) are typically used. Multiply these scales by a weighting factor to further delineate the importance of each criterion relative to the others. Assign criterion with low importance a value of 1; then assign criterion that is twice as important, for example, a value of 2.

   If the point scoring method (e.g., a total of 100 or 1000 points for all criteria) is used, allocate points to each criterion on the basis of its relative importance.

6. Ensure that all rating scales are aligned so that all favorable ratings are on the high end of the scale (e.g., both low cost and high return on investment should have a high rating number).

7. Discuss and select a voting methodology: majority voting for a suggested value (e.g., suggested by the leader) or an average of votes cast. Everyone must participate in the voting.

8. Conduct voting, which is done by rows so that each criterion is evaluated uniformly for the various alternatives. If rating scales are used, rate each alternative on a scale of 1 to 5. If the point system is used, assign points up to the maximum for each criterion (i.e., assign the maximum criterion points to all alternatives that satisfy the criterion).

9. Discuss divergent voting within each criterion-alternative cell. Large differences in voting may indicate that consensus may be difficult to achieve.

10. Add votes by column and sum the values for each alternative. For the point scoring method, merely sum the points. For rating scales, multiply each rating value by the criterion weighting factor and sum the products. (In the example above, the first column total is $25 = 1 \times 5 + 3 \times 2 + 1 \times 4 + 1 \times 4 + 2 \times 3$.)

11. Discuss the results to ensure that consensus has been reached. Remember that the alternative with the highest vote is not necessarily the consensus choice.

IS/IS NOT STRATIFICATION

Description

The Is/Is Not Stratification technique is used to help teams separate or stratify information, data, issues, etc. into like categories while attempting to reach consensus. It is also used to promote discussion and test the team's position regarding the alternatives rather than make a decision. This tool has many forms, ranging from its use with List Reduction (next tool in this chapter) to situations in which it is used to help define project scope or conditions under which problems occur.

During List Reduction, the technique involves asking the question, "Is an item the same as another, or is it not the same?" In defining the scope of a project or in helping to isolate the conditions under which problems occur, the technique expands to a matrix of descriptive criteria for evaluating relationships of variables. In addition to these uses, it can also be used to help stratify data into like categories in order to better understand patterns and relationships.

Although Is/Is Not Stratification shares with Affinity Diagrams and List Reduction the characteristic of grouping like items, it has one significant difference: Is/Is Not uses multiple criteria in a structured format to group like items. The difference recognizes that the relationship between items is generally not one-dimensional.

Additionally, this tool helps to define the suspected conditions that cause a problem, for example, and the data collection requirements needed to verify the cause-and-effect relationship (see Chapter Nine: Analyzing Cause and Effect).

Key Points

• Requires that all members have a clear understanding of all items on the list to be prioritized

- Aids in quickly grouping items using simple majority voting

- Promotes discussion of divergent views and tests the team's feelings regarding relationships within categories

- Can be used with a single criterion or multiple criteria that define the relationship

- Uses multiple criteria in a matrix format to evaluate problem conditions and project scope or to stratify data during analysis

Typical Applications

- Separate or stratify information, data, issues, etc. into like categories

- Define project scope

- Define suspected causal conditions when problems occur

- Define data collection requirements needed to verify a cause-and-effect relationship

- Stratify data into like categories for easy analysis, including various graphic displays (e.g., Pareto and Run Charts and Scatter Diagram)

Example

Customer Complaints			
	Is	**Is Not**	**Conclusions/ Comments**
What:	Empty Shelves Sci-Fi Horror Action New Releases	Comedy Romance Musicals Foreign Documentary Children/Classics	Shortage of several categories
When:	Friday nights & Saturday	Sunday through Thursday	Shortage during peak times
Extent:	85% of titles rented	50% of titles rented	
Who:	Customers <30	Older customers	Younger customers dissatisfied

FIGURE 6.4   Is/Is Not Stratification: Video One Customer Complaints

Steps

This tool is often used with List Reduction (the next tool in this chapter) since it helps to combine identical items (i.e., is an item identical or not: step 2 of List Reduction) and to evaluate the criteria used to prioritize items (i.e., does an item fit the criteria or not: step 5 of List Reduction). For other uses, the steps are as follows:

1. Identify the problem or situation to be stratified: name of project needing scope definition, problem needing definition of suspected casual conditions, or data that need to be stratified.

2. Set up a three-column matrix on the right side of a flip chart or board. The number of rows depends on the number of criteria used. Allow sufficient space to identify the row criteria on the left side.

3. Identify matrix columns as "Is," "Is Not," and "Conclusions and/or Comments."

4. Explain the rules of Brainstorming. Encourage everyone to participate.

5. Discuss and select appropriate criteria for evaluation. These criteria usually relate to physical location, environmental conditions, time of occurrence (i.e., time of day, month, etc. or sequence), characteristics of the relationship (e.g., duration, frequency, or type of event), and individual or group involvement. Identify the rows of the matrix with the criteria selected.

6. Brainstorm conditions or characteristics that fit the matrix. If the condition or characteristic fits the criteria, include it in the "Is" column; if not, place it in the "Is Not" column. Conclusions or comments are placed in the right column of the appropriate row.

7. Items are evaluated and discussed on a criterion-by-criterion basis.

8. Discuss results.

LIST REDUCTION

Description

List Reduction is often the first technique used after idea generation to start the convergence process leading to consensus. The objective is to quickly prioritize the ideas into a manageable number of important items (i.e., ten or less) using a series of filters or selection criteria. List Reduction usually involves only simple evaluation criteria and majority voting. It is usually followed by Criteria Rating Forms, Paired Comparisons, or Weighted Voting.

Key Points

- Gives all members equal voice

- Quick and easy method to prioritize lists

- Requires that all members have a clear understanding of all items on the list

- Compares all items using simple evaluation criteria

- Eliminates duplicated items and groups very similar items

- Aids in prioritizing items into broad categories (e.g., important versus less important) using simple majority voting on an item-by-item basis (alternate technique: allocate fixed number of points to participants)

- Does not eliminate less important items, but holds them for later processing

Typical Applications

- Generally follows idea generation activities and precedes other convergence activities leading to consensus

- Prioritize, evaluate, and select problem areas or process improvement opportunities

- Prioritize, evaluate, and select solutions

Example

		Perceived Problems/Improvement Opportunities
1		[Low return on sales]
2	A	Long check-out time
2	A	Check-out process not efficient
3	B	Not enough titles
4		Store layout poor
5		[Insufficient display space]
	B	Wrong movies in stock
	B	[Too few science fiction movies]
	B	Too few copies of new releases
6		[Declining profits]
	B	Poor selection
		~~Too long to check out~~
7		[No drop box for after hours]
8		No computer controls

FIGURE 6.5 List Reduction: Video One Problems/Improvement Opportunities

Steps

1. Clarify all items before selecting evaluation criteria and List Reduction begins. The person who suggests an idea explains it. Other team member comments should be confined to clarifying the item rather than evaluating its merit. Encourage everyone to participate.

2. Combine identical items (i.e., cross out all duplicate items) and group closely related items (e.g., linking them with an alphabetic identifier such as group A, group B, etc.).

3. Number all remaining items for identification purposes.

4. Identify and select the evaluation criteria, including the order in which they will be used. It is preferable to limit the criteria to questions that have a yes or no answer.

5. Criteria for evaluating, or prioritizing, the list might use the following questions:

   • Is it a critical problem or opportunity that needs to be solved or pursued?

   • Is the item within the control and influence of the team?

   • Are the resources available?

   • Is the payback sufficient?

6. Limit voting to a simple majority (i.e., one half the members plus one). Items that do not receive a simple majority should be bracketed (i.e., [ ]) for possible future consideration. (Variant technique allocates votes to each member equal to the total number of items, divided by three (e.g., 4 votes if there are 12 items). Members vote a maximum of one vote per item. Items receiving few votes are bracketed.)

7. Items are evaluated in order, and those without brackets continue into the next round of evaluation using the next evaluation criterion.

8. Continue prioritizing the list using additional, more stringent evaluation criteria or filters until 10 or fewer items are on the list. Use only one criterion for each iteration.

9. Briefly discuss the bracketed items to ensure that an important but unpopular alternative has not been set aside.

10. Proceed to another consensus tool: Criteria Rating Form, Paired Comparisons, or Weighted Voting.

PAIRED COMPARISONS

Description

Paired Comparisons help teams to identify preferences for various alternatives in their attempt to reach consensus. The tool allows all team members to vote for their preference in a series of Paired Comparisons of alternatives based on a simple evaluation criterion. This voting tests the team's position regarding the alternatives and helps to promote discussion rather than make a decision.

This tool is generally used when the list of alternatives has been reduced to a manageable level (e.g., six or less) as an alternative to Criteria Rating Forms or Weighted Voting. A similar but more analytical method of comparing pairs of alternatives is provided under Prioritization Matrices (Chapter Ten). That technique uses numerical ratings to indicate the strength of the preference of one item versus its paired alternatives.

Key Points

- Gives all members equal voice
- Comparison of pairs simplifies the ranking of a list
- Requires that all members have a clear understanding of all alternatives
- Compares pairs of alternatives using simple evaluation criteria
- No formal decision criteria
- Forces members to make decisions about alternatives that are closely ranked

Typical Applications

- Prioritize, evaluate, and select problem areas or process improvement opportunities
- Prioritize, evaluate, and select solutions
- Generally used after list of alternatives has been reduced using List Reduction

Example

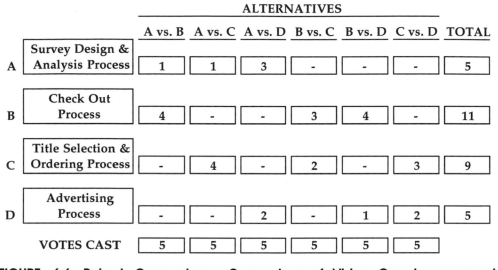

		A vs. B	A vs. C	A vs. D	B vs. C	B vs. D	C vs. D	TOTAL
A	Survey Design & Analysis Process	1	1	3	-	-	-	5
B	Check Out Process	4	-	-	3	4	-	11
C	Title Selection & Ordering Process	-	4	-	2	-	3	9
D	Advertising Process	-	-	2	-	1	2	5
	VOTES CAST	5	5	5	5	5	5	

FIGURE 6.6 Paired Comparisons: Comparison of Video One Improvement Opportunities

Steps

1. Clarify all items before selecting evaluation criteria and Paired Comparisons begins. Encourage everyone to participate.

2. Identify and select the evaluation criteria, including the order in which they will be used. Several iterations of the Paired Comparisons can occur, depending on the number of criteria selected. Use only one criterion at a time.

3. Criteria for evaluating pairs of alternatives might address the following questions:

    • Is it a critical problem or opportunity that needs to be solved or pursued?

    • Is the item within the control and influence of the team?

    • Are the resources available?

    • Is the payback sufficient?

4. Draw a grid with alternatives listed as row headings and combinations of paired alternatives as column headings. Code the alternatives (i.e., numerical or alphabetical codes) to simplify the headings for the pairs. The number of alternatives determines the number of pairs. If there are six alternatives (N = 6), for example, there will be 15 pairs [(N × N – 1)/ 2 = 15].

5. Voting is conducted on a column-by-column basis. Each member must vote for one of the alternatives in the pair based on the agreed-upon criteria.

6. Row totals signify the preference of the team based on the criteria evaluated. Column totals signify the number of votes cast, which should equal the number of participants.

7. Discuss the results to ensure that consensus has been reached. Discuss extreme voting behavior. Remember that the alternative with the highest vote is not necessarily the consensus choice.

WEIGHTED VOTING

Description

Weighted Voting helps teams to identify preferences for various alternatives in their attempt to reach consensus. The tool allows all team members to vote for their preference. This voting tests the team's position regarding the alternatives and helps to promote discussion rather than make a decision.

This tool is generally used when the list of alternatives has been reduced to a manageable level (e.g., ten or less) and can be used as an alternative to Criteria Rating Forms or Paired Comparisons.

Key Points

- Gives all members equal voice
- Fast method to select alternatives
- Requires that all members have a clear understanding of all alternatives
- Compares alternatives using simple evaluation criteria
- No formal decision criteria

Typical Applications

- Prioritize, evaluate, and select problem areas or process improvement opportunities
- Prioritize, evaluate, and select solutions
- Generally used after list of alternatives has been reduced using List Reduction

Example

TEAM MEMBER	ALTERNATIVES				
	Survey Design & Analysis Process	Check Out Process	Title Selection & Ordering Process	Advertising Process	VOTES CAST
1	0	6	0	0	6
2	0	0	6	0	6
3	0	3	3	0	6
4	4	1	1	0	6
5	0	2	0	4	6
TOTAL	4	12	10	4	

FIGURE 6.7   Weighted Voting: Voting for Video One Improvement Opportunities

Steps

1. Clarify all items before selecting evaluation criteria and Weighted Voting begins. Encourage everyone to participate.

2. Identify and select the evaluation criteria, including the order in which they will be used. Several iterations of Weighted Voting can occur, depending on the number of criteria selected. Each iteration generally uses only one criterion for evaluation.

3. Criteria for evaluating alternatives might address the following questions:

   • Is it a critical problem or opportunity that needs to be solved or pursued?

   • Is the item within the control and influence of the team?

   • Are the resources available?

   • Is the payback sufficient?

4. Draw a grid with the alternatives listed as column headings and team member names as row headings.

5. Give each member one-and-a-half points for each alternative (e.g., six alternatives would yield nine votes). Encourage members to spread their votes, but they may allocate as many points as they choose to any particular alternative.

6. Members decide their votes before the results are tallied (preferably recorded on paper). Votes are then recorded by alternative, not by member.

7. Discuss extreme voting behavior.

8. Discuss the results to ensure that consensus has been reached. Remember that the alternative with the highest vote is not necessarily the consensus choice.

CHAPTER SEVEN

PROCESS DEFINITION

Process definition begins with the identification of process boundaries, various characteristics of the process, and baseline performance measures. This is often followed by graphic representations using Flowcharts. Finally, documentation in the form of standard operating procedures is created where appropriate (e.g., for auditing purposes or for ISO 9000 registration). The first two tasks are addressed in this book, but standard operating procedures are not.

Three tools and techniques are included in this chapter:

- Flowcharts

- Process Analysis Worksheet

- Tree Diagram

The Process Analysis Worksheet is the starting point for process definition (see Step 2 of the Quality Improvement Model in Chapter Four). Flowcharting the process is next using a Top-Down Flowchart, then a Deployment Flowchart, and/or possibly a Tree Diagram or Work Flow Diagram (a spacial Flowchart) for certain detailed elements of the process.

The Process Analysis Worksheet establishes the process boundaries, describes the customer–supplier relationship in measurable terms, and provides a means for summarizing baseline measures of current performance and gaps. This worksheet helps to define customer needs and expectations, supplier specifications, and input specifications and identifies participants in the chain of customers and suppliers. Key elements of the data collected can be summarized in a Table of Performance Indicators and Drivers (introduced in Chapter Three).

Flowcharts come in a variety of different configurations, but they generally include inputs, outputs, and the sequence of activities. Additional details can be included, such as responsibility for activities, costs and value of each step, and the

81

time for each activity. These details help in understanding customer–supplier relationships, determining non-value steps that should be eliminated, and deciding where additional data need to be collected in order to eliminate process variation.

Review of baseline information in Flowcharts and the Process Analysis Worksheet is often sufficient for process stakeholders and owners to identify potential improvements. Obvious deficiencies can be spotted and corrected, but significant improvement usually requires that the three "V's" of process improvement be addressed: variation in the process, non-value activities, and velocity (i.e., cycle time).

Finally, the search for cause-and-effect relationships and the validation of root causes associated with the three "V's" of process improvement complete the diagnosis of a process that starts here (see Step 3 of the Quality Improvement Model in Chapter Four).

The Tree Diagram is primarily used for planning purposes, but it can be applied in process definition, which is why it is included in this chapter. The most common process definition use for a Tree Diagram documents the steps to assemble parts (or service actions) into subassemblies (or service bundles) and final assembly (or service program). Additionally, it can be used to describe a decision process, also known as decision trees. Another use is in cause-and-effect analysis with the Five Whys technique (Chapter Nine).

Other planning tools, such as the Activity Network Diagram (Chapter Eleven), can also be used in defining processes. The level of detail required for this tool, however, is beyond the needs of most process definition efforts.

FLOWCHARTS

Description

Flowcharts are graphic representations used to define processes. The charts range from the macro level (such as Top-Down) to the micro level (e.g., Deployment and Work Flow Diagrams) in showing various levels of detail. Flowcharts are generally used with other tools and techniques, such as the Process Analysis Worksheet. At a minimum, they show the sequence of steps and often include:

- Types of operations (activities, decisions, documents, and communication)
- Input and output documents, products, or services
- Responsibility, geographic location, and customer–supplier relationships
- Elapsed and process time (actual and theoretical estimates)
- Costs and value added
- Potential data requirements and quality improvement opportunities

An example and brief description of each of the primary Flowcharts follows. The Deployment Flowchart tool is described in greater detail because it is at a level of detail necessary to begin process improvement.

Macro Flowchart

Macro Flowcharts show the sequence of major activities in a process (or planning steps of a project), which are often done before attempting micro-level Flowcharts. These charts include all the major steps in the process, which are described at a very summary level. They are constructed with the sequence of process steps shown in either columns or rows, starting with the first step in the top left corner. If the steps flow in columns, then steps at the bottom of one column are connected to steps at the top of the next column. Likewise with row charts, steps at the right margin connect with the next step below on the left margin. Because these charts show the sequence of the major process steps, they help teams gain a perspective of the total process.

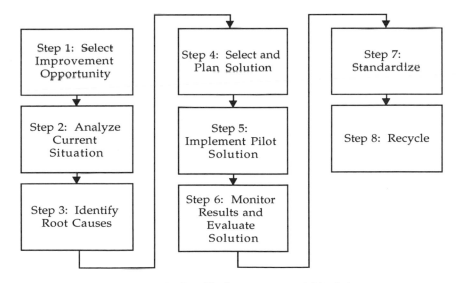

FIGURE 7.1   Macro Flowchart: Quality Improvement Model

Top-Down Flowchart

Top-Down Flowcharts show key steps in the process (or planning steps of a project). Key steps are listed in order across the top of the page, starting with the first step on the left. Activities necessary to achieve key steps are listed under the key steps. Both the number of key steps and the number of activities are usually limited to seven in order to focus on the vital few activities. Only steps that add value are included; inspection and rework are omitted. Comparison of Top-Down Flowcharts with more detailed Flowcharts highlights non-value-added steps. (See the Process Analysis Worksheet for additional details on completing a Top-Down Flowchart.)

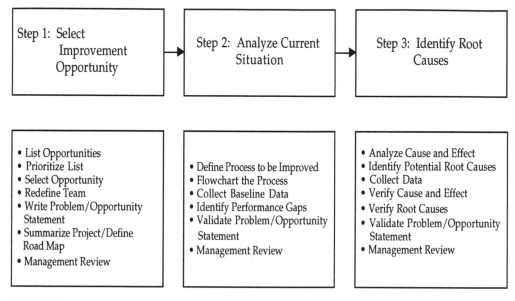

FIGURE 7.2   Top-Down Flowchart: Quality Improvement Model

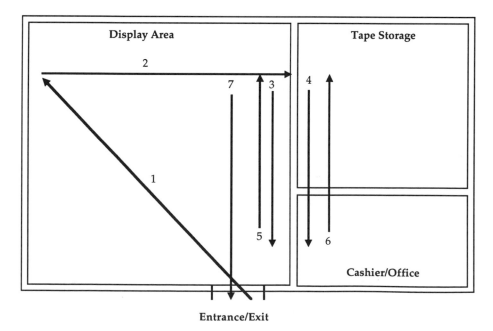

FIGURE 7.3   Work Flow Diagram: Video One Check-Out Process

Work Flow Diagram

Work Flow Diagrams graphically display the movement of people, work, or information in a process. The chart is created by tracing movements on a floor plan or other representation of a geographic location. This tool is helpful in illustrating illogical flows of work or information and in demonstrating the need to co-locate common activities. The chart can be constructed to show not only the patterns of interaction and the work or information flows, but the sequence of activities as well. In conjunction with other detailed Flowcharts, Work Flow Diagrams can help to illustrate the need to change the sequence of steps and/or the physical arrangement of facilities.

Deployment Flowchart

Deployment Flowcharts are a method of charting work processes to identify not only the sequence of work activities and decisions, but also the responsibility (i.e., individual or organization) and/or location for performing tasks and making decisions. In addition, Deployment Flowcharts illustrate interaction between individuals or organizations (customer–supplier relationships). This interaction is usually in the form of communication or other involvement (e.g., document review and meetings). Elapsed time from the first step and process time for each step are generally indicated in the left column, which is the time line. Alternatively, these charts can be used to evaluate value-added process time versus theoretical process time.

Key Points

- Simplifies understanding of a process
- Relates activities by responsibility and location
- Capable of tracking total elapsed time and identifying process time (e.g., actual, value added, and theoretical)
- Horizontal linkages represent supplier–customer relationships between organizations
- Vertical linkages show time sequence and customer–supplier relationships within the organization
- Highlights sequential work and information flows that could possibly be replaced by parallel, or concurrent, activity
- Highlights failure points that require rework and corrective action, a primary source of variation and non-value steps
- Illustrates multiple process stakeholders and the need for a process owner

- Helps to identify bottlenecks and critical process steps (tasks and decisions) that require additional data for process improvement

Typical Applications

- Understand the sequence of and responsibility for performing activities within a work process
- Establish a baseline for process improvement and process redesign
- Identify specific areas where additional data are needed
- Identify the steps in planning a project

Symbols

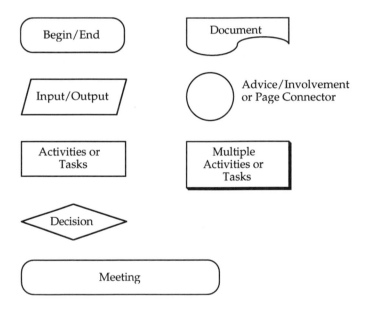

FIGURE 7.4   Flowchart Symbols

Example

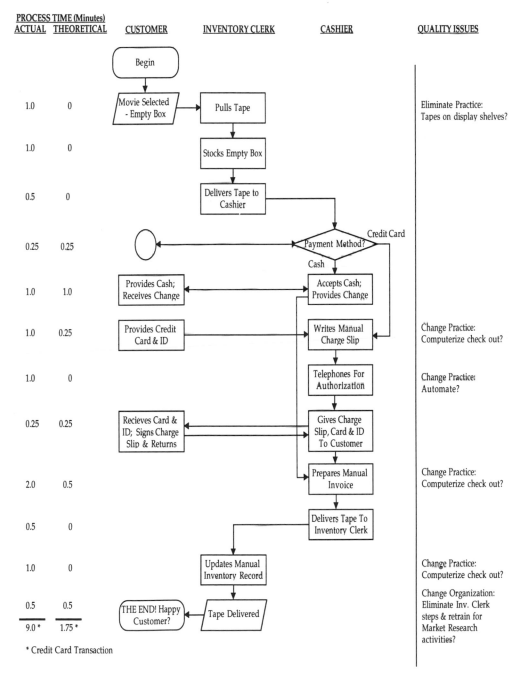

FIGURE 7.5   Deployment Flowchart: Video One Check-Out Process

Steps

1. Describe the process, including its boundaries and any assumptions. Use the Process Analysis Worksheet to describe outputs and customer–supplier relationships.

2. Brainstorm the steps in the process (i.e., activities, decisions, meetings, communication, inputs, and outputs). Initially they may be out of order. Describe activities in verb-noun format and outputs in noun-verb format. Use different colored 3 × 5 Post-It™ notes to identify the different types of steps (i.e., activities, decisions, etc.) or draw the symbol on the same color 3 × 5 Post-It™ notes.

3. Identify who is responsible for the activity or where each activity is performed on each Post-It™ note. Also list other organizations, individuals, or locations involved in each step.

4. After all items are written, label columns on a large sheet of paper (e.g., butcher paper or flip chart paper) for each organization, individual, or location involved. Allow space on the left for the time line and on the right for quality issues. Column widths should be as wide as the Post-It™ notes. (Alternatively, the Flowchart can be displayed horizontally, where all column instructions become row instructions.)

5. Arrange the notes in the general sequence of the process and in the appropriate columns. Actions that occur simultaneously should be on the same horizontal line and subsequent steps are arranged below. (Several alternate approaches exist, including backward chaining, which starts with the output delivered to the customer and chains backward to the input. Another approach starts with an activity in the middle of the process and then places activities before (above) or after (below) this middle task. Inputs and outputs are then linked with appropriate activities.)

6. After an initial review, transfer the information, with symbols, to the butcher paper or flip chart while removing the notes. Next, add connecting lines and arrows to show the process flow. Ensure that symbols are drawn to indicate involvement of other people or organizations in each step. Responsibility for a meeting or verbal communication is denoted by the activity symbol in the column of the organization or individual responsible for convening the meeting or initiating the communication. Show the meeting or advice/involvement symbol in the columns of others involved.

7. Add the time line next to indicate elapsed time from the beginning of the process, actual process time, and/or estimated theoretical process time for each step. The actual and theoretical process times for each step highlight waste in the process and are the most commonly used. The theoretical time reflects activities that add value; an activity that adds no value receives a "zero" theoretical time.

8. All process stakeholders and the process owner should review the draft Flowchart for accuracy and make recommendations for revisions.

9. Finally, quality improvement issues should be noted in the quality issue column. The following areas should be explored:

   - Steps that add little value with long actual process times

   - Sequential work and information flows rather than parallel, or concurrent, activity; illogical work and information flows

   - Failure points (that require rework and corrective action, a primary source of variation and non-value tasks), bottlenecks, and critical process steps (tasks and decisions) that require additional data for process improvement

   - Customer–supplier relationships that may not be clearly defined

   - Manual steps between computer systems, duplicate tasks, multiple approvals, and excessive process checks and inspection

   - Potential elimination of functional or departmental boundaries to reduce delays, mistakes, and unnecessary specialization

PROCESS ANALYSIS WORKSHEET

Description

Process Analysis Worksheets are the starting point for Process Definition (Step 2 of the Quality Improvement Model). These worksheets help to define process boundaries while describing customer–supplier relationships in measurable terms. They also provide a means for summarizing baseline measures of current performance and gaps.

Used with Deployment Flowcharts, these worksheets provide sufficient baseline information for teams to begin diagnosis of process deficiencies. This ultimately leads to the search for cause-and-effect relationships and the validation of root causes. Summary data are used to build a Table of Performance Indicators and Drivers (Chapter Three) that illustrates the changes in performance measures while highlighting the cause-and-effect relationship between drivers and outcomes.

Key Points

- Identifies process outputs, customers, and customer needs and expectations in order of priority

- Translates customer requirements into measurable supplier specifications

- Provide a means for comparing current performance with customer expectations

- Identifies process inputs, suppliers, input specifications, and input performance measures

- Used with Flowcharts and the Table of Performance Indicators and Drivers to create baseline process definition

Typical Applications

- Process definition during process improvement

- Planning for new products and services

Example

Process Analysis Worksheet

Output: **Tape Delivered to Customer**

Primary Customer: **Rental Customer (E) End User**

Secondary Customer(s): **Customer's Family (E) End User**

Customer-Supplier Data:

Customer Requirements	Ranking	Supplier Specifications	Current Performance	Performance Gaps
Large Selection of Titles	1	Number of titles	1000	?
		Number of new titles	200	?
		Copies per title	2.5	?
Fast Service	2	Check out time	6 to 10 minutes	3 to 7 minutes
Convenient Service	3	Convenient location	?	?
		Long hours	10 hours (10 AM - 8 PM)	?
		After hours drop box	No	?
Low Price	4	Less than $2/rental	$2.10 average price	?

FIGURE 7.6   Process Analysis Worksheet: Customer–Supplier Data

Process Analysis Worksheet

Output: Tape Delivered to Customer

Primary Customer: Rental Customer (E) End User

Input Data:

Inputs	Input Suppliers	Input Specifications	Current Performance	Performance Gaps
Rental Tapes Supplied	ABC Distribution	Unlimited new titles as released	?	?
		24 hour delivery	?	?
		Competitive prices	?	?

Staff Trained:
  Manager
  Cashier
  Inventory Clerk
  Accountant
  4 Others
	Geraldo
	Ernestine
	Willard
	Ivan
	Other employees
		Cross-training:
		  Check out process
		  Inventory control
		  Credit card approval
		  Invoicing
			No training	4 hours per month

Cash Register Receipt Produced	NCR	Error free operation	?	?
Credit Card Authorization Made	Visa/Master Card	Fast, manual authorization	1 minute	?

FIGURE 7.7   Process Analysis Worksheet: Input Data

Process Analysis Worksheet

Process Name: Check Out Process

Process Scope: (From)   Customer Selection of Tape   (To)   Tape Delivered to Customer

Tops Down Flowchart:

Top Level:	Customer selects empty movie box	Inventory clerk swaps box for tape	Cashier processes payment	Cashier completes invoice	Inventory clerk updates inventory records	Deliver tape to customer
Activity List:	Movie selected	Pulls tape Stocks empty box Delivers tape to cashier	Requests cash or credit card Takes cash and makes change or Writes manual charge card slip Telephones for credit card authorization Obtains customer signature	Prepares manual invoice Delivers tape to inventory clerk	Updates manual inventory record	Tape delivered to customer

FIGURE 7.8   Process Analysis Worksheet: Top-Down Flowchart

Process: Check Out

Output: Tape Delivered to Customer

	"As Is"	"Desired State"
PERFORMANCE INDICATORS:		
Outcomes		
Customer Satisfaction/Loyalty		
Satisfied Customers	5%	100%
Dissatisfied Customers	85%	0%
Process Variables & Attributes		
Check Out Time (minutes)	6 to 10	
Six Months		3 to 5
One Year		3
Customers per Hour		
Peak Period	?	?
Off Peak	?	?
Titles per Customer	?	?
Cash Register Failures per Day	?	?
Percentage Credit Cards	?	?
PERFORMANCE DRIVERS:		
Sub-Processes		
Inventory Management Process	see flowchart	see below
Credit Card Process	see flowchart	see below
Invoicing Process	see flowchart	see below
Practices		
Credit Card Authorization	manual	automated?
Inventory Record Keeping	manual	automated?
Invoicing	manual	automated?
Tape Inventory Storage	separate room	on shelves?
Structural Factors		
Cross Training per Employee	none	4 hours/mo.

(Bracket label at left spanning Performance Indicators section: "Family of Measures")

FIGURE 7.9   Process Analysis Worksheet: Performance Indicators and Drivers

Customer–Supplier Data

This worksheet consists of the following elements (Figure 7.6):

Output

Enter the output statement (noun-verb format) that describes the product produced or service provided. Your work unit is the supplier of this output.

Checklist:

- Specific product or service needed or expected by customer?
- Neither too broad nor narrow and described clearly?
- Not an outcome or result, such as a satisfied customer?
- Not a step in a process?

Primary Customer

Identify the next individual, work unit, or external company receiving the output. This is the primary customer. Indicate if customer is internal (i.e., different work unit in your company) or external (i.e., different company). Use an "(I)" for internal and an "(E)" for external customers. Identify all end-users. If you have identified yourself as a customer, you have probably defined a step in the process rather than an output.

Checklist:

- Specified by name?
- Not your manager, unless he or she adds value (i.e., co-supplier)?
- End-users identified?

Secondary Customer(s)

Add the names of the next individual or organization (i.e., internal work unit or external customer) in the process after the primary customer. Also indicate whether these secondary customers are internal or external, as above. Identify the sequence of these customers if known (i.e., in order of receiving the output). If one of the secondary customers is an end-user, identify as such.

Checklist:

- Specified by name?
- Not your manager, unless he or she adds value (i.e., co-supplier)?
- Identified sequence of customers and end-users?

Customer Requirements

List primary customer needs and expectations for the output. Verify these requirements with primary customers and uncover additional requirements of secondary customers. Categorize requirements using qualitative and quantitative terms: quality (e.g., reliability, accuracy, and conformance to specific dimensions), cost, and delivery (e.g., quantity, timeliness, and completeness). Specific requirements should be attainable, measurable, well communicated, and mutually agreed upon.

Not all requirements need to be specified, for example, where implied requirements are understood. When customers specify needs and expectations that are generally taken for granted, it usually means that another supplier has not met these expectations.

Customers are responsible for creating an environment in which suppliers are encouraged to ask for better and more precise requirements. The supplier often knows more about the requirements than the customer and should take the lead in such situations. In some cases, requirements are non-negotiable. Where conflicting requirements for multiple customers exist, the supplier is generally in the best position to reconcile differences.

In the Video One example in Figure 7.6, the team identified "Large Selection of Titles" and "Low Price" as key customer requirements. The team realizes that these requirements are important but do not relate directly to the check-out process and, therefore, will be analyzed later. The question marks indicate that some data were not known when the worksheet was prepared.

Checklist:

- Clear?
- Specific as possible?
- End-user requirements considered?
- Customer objectives known?
- Negotiations required?

Ranking

Identify primary customer's ranking of requirements (i.e., in order of importance). This helps to separate the non-negotiable and important requirements from less important ones. Additionally, it helps to resolve conflicting requirements.

Supplier Specifications

Translate customer requirements into supplier terminology. In some cases, the customer requirements and the supplier specifications are identical. Express supplier specifications in measurable terms.

Checklist:

- Related directly to customer requirements?
- Output described in supplier terminology?
- Measurable and objective?
- Realistic and attainable?

Current Performance

Add known performance relative to each supplier specification. Initially, these measurements may be lacking, but the collection of baseline data to analyze the process will uncover current performance. Collection of baseline data is critical to Step 2, Analyze the Current Situation, of the Quality Improvement Model. Focus data collection on important specifications (i.e., important customer requirements).

Checklist:

- Current performance against supplier specifications?
- Consistent, timely, and accurate data available?

Performance Gaps

Calculate the difference between supplier specifications and current performance. Large gaps for important specifications should receive priority attention. There are three critical times to measure performance: before (to measure the capability of the process), during (to measure if the process is producing acceptable outputs), and after (to determine if acceptable outputs were delivered to customers). Select appropriate and consistent times to collect data.

Checklist:

- Large gaps for important specifications identified?
- Gaps identified before production (i.e., prevention), during production (i.e., inspection), or after shipment (i.e., failure)?

Input Data

This worksheet consists of the following elements (Figure 7.7):

Inputs

List inputs received by your work unit that are necessary to produce the output. Include various tangible resources (e.g., people, equipment, facilities, capital funds, material resources) as well as intangible resources, such as training and knowledge. Your work unit is the primary customer for these inputs, which are your suppliers' outputs. Describe them in noun-verb format. Change verbs to reflect your receipt of an output.

In the Video One example in Figure 7.7, the team identified "Rental Tapes Supplied" as a key input. Later in the analysis, the team realizes that this input is important but does not need to be analyzed immediately because they will address the check-out process first. The question marks indicate that some data were not known when the worksheet was prepared.

Checklist:

- Specific product or service needed or expected by you?

- Neither too broad nor narrow, described clearly?

- Not a step in a process?

Input Suppliers

Add the names of the individual or organization (i.e., internal departments or external company) that supplies the input to you. Also indicate whether the supplier is internal or external.

Checklist:

- Specified by name?

- Not within your work unit, indicating too narrow a process scope?

Input Specifications

Define input specifications in the same manner as supplier–customer negotiations described above. Your work unit is the primary customer for its supplier. (Treat your suppliers as you would like your customers to treat you!) The supplier specifications negotiated with your customers influence your requirements, or input specifications, with your suppliers.

Define input specifications in terms of customer requirements where possible.

Checklist:

- Clear?

- Specific as possible?

- End-user requirements considered?

- Negotiations required?

Current Performance

Add current performance for each input specification if known. Initially, these measurements may be lacking, but the collection of baseline data to analyze the process will uncover these data. Data collection should focus on important input specifications, which usually correspond with important customer requirements.

Checklist:

- Current performance measured against input specifications?
- Consistent, timely, and accurate data available?

Performance Gaps

Calculate the difference between specifications and current performance. Large gaps for important specifications should receive priority attention. Flawed inputs reflect failures in prior steps.

Checklist:

- Large gaps for important specifications identified?

Top-Down Flowchart

This worksheet consists of the following elements (Figure 7.8):

Process Name and Scope

Enter a name for the process being analyzed. Enter the first and last major steps in the process that define its scope.

In the Video One example in Figure 7.8, the team has already decided to address the check-out process, and their first attempt to flowchart the process used a Top-Down Flowchart.

Top Level

List the major steps required to produce the output, starting with the first one on the left side. Add major steps to the right of the first one until the output is identified on the right side. The first and last steps define the scope of the process. Limit the number of steps to seven to ensure that the view of the process begins at the summary level.

Activity List

List all significant activities performed under each top-level step. Listing activities in approximate order helps in completing the Deployment Flowchart.

Checklist

- Do procedures exist that document the process?
- Do the steps cover major activities, milestones, and measurement points?
- Do the activities reflect an adequate level of detail to guide the work?

Deployment Flowchart

Complete a Deployment Flowchart (see details provided earlier in this chapter).

Checklist:

- Shows sequence of activities and decisions?
- Shows responsibilities?
- Shows resources as inputs?
- Clear enough to train a new person?
- Identifies issues: illogical work and data flows, non-value activities, and potential sources of variation?

Performance Indicators and Drivers

This table (Figure 7.9) highlights the vital few performance measures within the scope of the analysis. Furthermore, it is useful in summarizing data for presentations and documenting project activities.

Select and enter critical measures that relate to the "as is" condition and the "desired state" of the process (the improvement opportunity). Measurements should be derived from customer requirements, supplier specifications, and input specifications. Initially, only limited data may be available. As the analysis progresses through Step 3 of the Quality Improvement Model (Identify Root Causes), the data for this table should be available.

In addition to data related to inputs and outputs, this table includes:

- Outcomes, or results, achieved if outputs satisfy customer requirements
- Measures related to productivity: process efficiency of work (outputs divided by inputs) and process utilization (actual use of resources relative to potential use)
- Process activity levels and other process characteristics

More importantly, the table helps to focus attention on performance drivers in need of change before outcomes can improve. During process improvement, various steps, practices, and structural factors will be identified as root causes of process deficiencies, which must be addressed if the improvement opportunity is to be realized.

In the Video One example in Figure 7.9, the team identified the vital few measures to help them analyze the check-out process. The question marks indicate that some data were not known when the table was prepared. As the team continues its search for root causes, additional indicators and drivers may be added. Potential solutions (indicated with a question mark) shown in the "Desired State" column are also included.

Checklist:

- Critical few measures?

- Quantifiable measures?

- Easy to collect?

- Measurements related to customer requirements and key process steps?

- Measures related to process variation, non-value steps, and process velocity?

- Measures linked with desired outcomes?

TREE DIAGRAM

Description

The Tree Diagram (or systematic diagram) is an analytical and planning tool that is used to break down issues, statements, or ideas until actionable items are identified. Its primary use is for planning purposes, where it is used to detail the full range of paths and tasks that need to be accomplished to achieve a goal or subgoal.

Process definition applications of the Tree Diagram rely on its ability to organize expanding levels of dependent activities into a logical sequence. Applications include processes producing products (from parts and subassembly to final assembly) or services.

Tree Diagrams can also be used in identifying cause-and-effect relationships using the Five Whys technique (Chapter Nine: Analyzing Cause and Effect).

Key Points

- Identifies the sequence of tasks needed to address an issue

- Breaks down issues into components from which actionable items are identified and assigned

- Addresses the logical link between layers of detail, from the general to the specific

Typical Applications

- Identify the logical sequence of steps required to carry out a goal or project

- Understand the logical sequence in assembling a product or providing a service

- Identify all possible causes of a problem or alternatives for its solution

- Break down goals and objectives into specific, detailed actions

- Organize output of an Affinity Diagram (Chapter Six) for use with Matrix Diagrams and the Prioritization Matrix (Chapter Ten).

Example

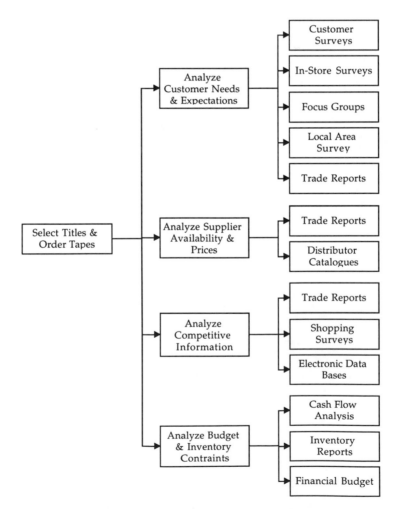

FIGURE 7.10 Tree Diagram: Video One Title Selection Process

Steps

1. Choose the issue, project, or goal to be analyzed and develop a goal statement (verb-subject format). Start at the broadest level of detail, which is placed on the left side.

   (If the root cause is sought using cause-and-effect analysis and the Five Whys technique, state the "as is" problem. If a product or service process is to be defined, then use the product or service as the goal.)

2. Identify the major headings in the second layer of detail. Use previously developed Affinity Diagrams (i.e., header cards) or Interrelationship Digraphs (i.e., cards leading into the goal statement) or Brainstorm new ideas. Use 3 × 5 Post-It™ notes or index cards for this and subsequent layers until all the items are arranged in a logical sequence.

3. Continue generating each successive layer of detail by identifying the tasks that must be accomplished. Do this by asking "In order to accomplish this, what must happen?" until all tasks are logically ordered.

   (When cause-and-effect analysis is performed with the Five Whys technique, each layer of detail is developed by asking "Why does this happen?" or "What causes this?" for each layer.)

4. Review the diagram to ensure that all tasks are included and that logical sequences have been established. Test the logic in going from the specific to the general (i.e., right to left) by asking "Will these items, actions, or causes lead to this result or effect?" Test the logic of the general to the specific (i.e., left to right) by asking "Will these results or effects occur if these items and actions occur or if these causes exist?"

CHAPTER EIGHT

COLLECTING DATA

Data collection occurs at various times during quality improvement. Data are often available in Step 1 of the Quality Improvement Model to help select the problem or improvement opportunity. In Step 2, where the current situation is analyzed, the process is defined using the Process Analysis Worksheet and Flowcharts that help to identify data collection requirements. As the team searches for root causes in Step 3, it will collect data to verify cause-and-effect relationships. This is often the most important data collection event in quality improvement.

Finally, in Steps 5 through 7 of the Quality Improvement Model, the team captures additional data to track results, first in the pilot and then when process improvement is being standardized throughout the organization. Here, data validate that process changes have had a measurable impact on the organization. All data collection activity, from baseline data to final results, should include measurement of desirable outcomes.

Before collecting data, the team needs to establish a plan that defines what and how much data are needed, where and how they should be collected, how long and by whom they should be collected, and what assumptions are needed. It is usually helpful when capturing data to identify not only what it is, but also to capture any information that might bias the data. The need to collect statistically significant, unbiased data is a common requirement in quality improvement, and the plan should comprehend this need. Finally, the plan needs to define the tools to be used.

There are several considerations in collecting data: purpose, ease, validity, reliability, and bias. The key questions regarding the purpose of data are why is the data needed, what will it tell you, and how will you use it? Because data collection is time consuming, it is important that it be limited to vital data only. The data requested should, however, be as complete as possible to ensure that key data are not missed, which could require costly follow-up action. Not all data will require real-time data collection; in some cases, existing documents, such as computer printouts, written reports, etc., may be used

The data plan should address the ease of collecting data. Be prepared to explain the purpose of collecting the data and provide instructions on how data are to be captured to those who will actually collect the data. Special forms often need to be designed to simplify the process. All abbreviations and other nomenclature need to be defined. Forms completed with sample data are often helpful. Trial data collection periods not only test the forms, but also test the ease of using the data in analysis, either through manual or computerized means. A well-designed survey document eases both data collection and tabulation.

Finally, the data collection plan needs to cover the potential validity, reliability, and biases in data collection. Poorly constructed Sampling plans and poorly worded instructions often lead to sampling errors and variations in collection methodology. Projects that require statistically relevant data should obtain a statistician's review of the data collection plan before beginning.

The data collection tools and techniques are:

- Check Sheet
- Focus Groups
- Sampling
- Surveys

CHECK SHEETS

Description

Check Sheets are used to track the frequency of specific events that occur during a sampling period. They provide a means to record data using a matrix format in which one variable is the sampling period and the other is the event. Comparisons of events with cost or other variables that stratify data are the primary alternatives to the basic event versus time comparison.

Check Sheets begin the process of translating opinions into facts for later analysis. Generally, data are collected as they occur but can often be collected after the fact.

Key Points

- Provides data on events and their relative frequency
- Can provide an easily understood snapshot of a process and its problems
- Often the first step in using data to prove or disprove opinions

- Event data are collected in various categories for comparison with periods of time, costs, or other measures

- Necessary to ensure that data collected is from homogeneous populations

Typical Applications

- Gather data on the number of occurrences of various issues, problems, inputs, outputs, or outcomes during specified time periods

- Record occurrences of a variety of activities (e.g., work, rework, inspection, or idle time) in a variety of categories (e.g., by machine, shift, or team) during a sampling time period

- Count errors or failures by type during a time period

Example

	Mon.	Tue.	Wed.	Thur.	Fri.	Sat.	Sun.	Total
Check Out Time	✓✓	✓✓✓	✓✓✓	✓✓✓✓✓ ✓✓✓	✓✓✓✓✓ ✓✓✓✓✓ ✓✓	✓✓✓✓✓ ✓✓✓✓✓ ✓✓✓✓	✓✓✓	45
No New Releases		✓✓			✓✓✓✓✓ ✓✓	✓✓✓✓✓ ✓✓✓	✓✓✓	20
Poor Title Selection		✓✓✓			✓✓✓✓	✓✓✓✓✓ ✓✓✓		15
Can't Find Movies				✓✓	✓✓✓	✓✓	✓	8
Other		✓	✓	✓✓			✓✓	6
	2	9	4	12	26	32	9	94

FIGURE 8.1   Check Sheet: Video One Customer Complaints

Steps

1. Identify all categories and classifications of events to be recorded. Include an "other" category to capture data that do not fit the primary categories. Decide how long the data will be collected. Then determine the comparison measure: period of time (e.g., hour, shift, day, month, time since last occurrence, etc.), cost ranges, or other measures. Finally, if the population

is not homogeneous, it should be stratified (i.e., grouped) into uniform categories.

2. Create a Check Sheet form with clearly labeled rows and columns. Provide sufficient space for the check marks.

3. Ensure that everyone using the Check Sheet understands what events are being observed and how to record an occurrence. Additionally, make sure that everyone has sufficient time to complete the Check Sheet and understands the duration of data collection.

4. Consolidate results from multiple Check Sheets (where applicable). Determine how to present the data (e.g., Pareto Chart or other charts, see Chapter Ten: Analyzing and Displaying Data).

5. Determine additional data gathering requirements.

FOCUS GROUPS

Description

A Focus Group is an interviewing technique used to gather data from groups of customers. It offers suppliers an opportunity to better understand customer needs and expectations and to obtain constructive criticism in order to improve products and services. The data collected often include specific measures of customer satisfaction on various dimensions of product and service quality.

Focus Groups usually include a limited number of participants who represent a cross-section of customers and act as surrogates for the entire population of actual or potential customers. Participants are a small sample of a larger population. Focus Groups require a trained facilitator familiar with group interviewing techniques. The sessions are generally recorded with participant permission. Incentives to participate are common and range from free or discounted products and services to summary reports of group comments and recommendations.

In some cases, participants are reluctant to provide candid feedback in an open forum. This dilemma can often be resolved by using the Crawford Slip Method (Chapter Five), which protects the anonymity of participants.

Key Points

- Used to obtain knowledge about customer needs and expectations

- Provides constructive feedback from customers on both the need to improve product and services and suggested methods to accomplish these improvements

- Provides specific customer satisfaction data along various dimensions of quality

- Participants represent a cross-section of actual or potential customers

Typical Applications

- Gain knowledge of customer needs and expectations

- Measure customer satisfaction

- Assist Quality Improvement Teams to obtain customer feedback on improving processes that produce products and services used by participants

- Provide credible customer data to support quality improvement

Steps

1. Develop list of questions and arrange in a logical order using the three basic elements of a Survey: introduction, body, and conclusion (see the next section in this chapter, Surveys, and Chapter Fourteen: Question-naires). The introduction includes the purpose of the Focus Group. The body covers the main questions, starting with easy, non-threatening ones and moving to more sensitive ones. The conclusion includes the demographic questions.

2. Design a plan for analyzing results, particularly if the Crawford Slip Method will be used.

3. Select participants who reflect an accurate, fair cross-section of customers to ensure the validity of the Focus Group responses. Limit the size of the group to 12 participants.

4. Select a facilitator who is familiar with group interview techniques and determine meeting logistics.

5. Transmit participant instructions and logistics information. Exclude any information that might bias responses or affect willingness to participate. Include the following information:

   - Meeting logistics: location, date, time, and other details

   - Agenda that shows purpose, desired outcomes, and topics

   - Recording release form

   - Copy of the questionnaire, if appropriate

6. Provide appropriate incentives at the beginning of the meeting and state when participants should expect a summary report if offered. Consider

offering the incentive at the end of the session if it could potentially bias participant responses. Incentives should be geared to the level of effort required by participants and should represent appreciation rather than compensation.

7. Conduct group session. Record the session (audio or video) with participant permission.

8. Prepare report documenting key findings and recommendations. Distribute to participants, if offered.

SAMPLING

Description

Sampling is used to obtain information about a large group (the population) from a smaller representative group (the sample). The sample reflects the characteristics of the larger population and provides a cost-effective means of collecting data for decision making. The sample may be drawn from a specific lot (a finite population) or a continuous process (an infinite population).

Effective Sampling provides information that satisfies four key criteria: valid, timely, reliable, and economical. Sampling methods and the size of the sample often determine whether these criteria are satisfied. Types of Sampling include random, systematic, stratified, cluster, and two-stage.

Key Points

- Allows conclusions to be made about a large group from a much smaller one (i.e., inductive analysis where conclusions about a population are made from a small sample)

- Recognizes that collected data are not all the same and that some level of variation in results is expected

- Sample size is determined by the degree of precision required

- Most, but not all, populations fit the normal distribution or bell-shaped curve

Typical Applications

- Estimate the distribution of total defects, errors, or other failures in products or services

- Estimate the average number of defects, errors, or other failures produced during a given time period

- Estimate various process characteristics

- Compare performance statistics with process specifications

- Determine if a process is in control

Sampling Types

Simple Random Sampling

Random sampling is the most common and most powerful method of Sampling. In this method, there is an equal probability that any unit of the population will be part of the sample. To determine which units are to be sampled, random numbers are required. Random numbers can be obtained using published texts of random numbers, by the throw of the die, via a lottery system, or by using computer-generated numbers.

Obtaining true randomness in real-world samples can be difficult, expensive, or even impossible. Examples where random sampling **would not** work well include Sampling in a warehouse (where the chosen random sample may be buried beneath tons of other production) and Sampling theater audience reaction to a movie (the random sample could be a one-year-old child). Examples of random sampling include:

- Reviewing documents #1, 2, 4, 5, 13, and 14 (obtained from a random number generator) for errors

- Taking samples at 1:42 p.m., 1:53 p.m., 2:15 p.m., 2:31 p.m., and 3:22 p.m. (obtained as number of minutes after 1:30 p.m. from a random number generator) to measure output statistics

Systematic Sampling

Because simple random sampling from a population is frequently difficult, it is often easier to sample following a specific fixed pattern or time interval. This variation of random sampling is referred to as systematic sampling. Examples include taking every tenth product or service as a sample or taking a sample every 15 minutes.

Stratified Sampling

Stratified sampling occurs when known differences or categories exist within the population. Random sample(s) are taken from each category in the population to proportionally represent that category. An example of this type of Sampling would be to proportionally sample each job classification in an office in order to determine the number of customer contacts made per day. The closer the homo-geneity of each category, the more precise the overall samples will be.

Cluster Sampling

Cluster sampling is used only when a population is known to be relatively stable (i.e., without major variations). In this case, one subgroup of the population is used as the sample. Examples of this type of Sampling include selecting work orders processed between 8:00 and 9:00 a.m. as representative of an entire work day or selecting a 10:15 a.m. assembly line sample as representative of 8 hours of production.

Two-Stage Sampling

Two-stage sampling is commonly used in factories with large production num-bers. In this Sampling method, samples are first taken from the large population (typically by the random or systematic method). These samples, referred to as primary samples, are then sampled (typically randomly) in order to obtain a smaller sample group, referred to as a secondary sample. The secondary sample is then designated as representative of the original large population.

Sample Size

Determining the sample size and the number of samples that should be taken is affected by many factors. Some of these considerations include:

- Consistency of the process
- Difficulty of gathering data
- Cost of gathering data
- Intuition
- Relative importance or consequences of making an error about the popu-lation based on a sample
- Severity or frequency of the problem for which samples are being taken
- Size of the total population

It is beyond the scope of this book to review the varying levels of Sampling and the risk of error associated with each. Users who need additional information on this topic are referred to Appendix B: References and Resources.

SURVEYS

Description

Surveys (either in writing as Questionnaires or verbally as interviews) are means of gathering data. The data range from opinions, feelings, and outcomes (e.g., level of satisfaction) to specific data on products, services, processes, and practices. Surveys require the development of questions and a methodology for collecting the data. These methods include mail Surveys, personal or telephone interviews, and in-depth Questionnaires used in Benchmarking, for example (Chapter Fourteen: Questionnaires).

Questionnaires are usually completed by respondents in writing or used as the script for interviews (either in person or via telephone). Surveys involve only a sample of the population, whereas a census involves the entire population.

Key Points

- Planning for data needs, sources, and methods critical to success

- Surveys gather limited data from many respondents quickly at low cost

- Respondents to Surveys may be different from those who do not respond; response rates less than 25% are typically negatively skewed

- Interviews provide more in-depth information, but take longer and cost more

- Three parts to a Questionnaire, each with a specific purpose: introduction, body, and conclusion

- Two basic types of questions: unstructured (open-ended and fill in) and structured (yes-no, multiple choice, and scaled response)

Typical Applications

- Determine customer needs, expectations, complaints, suggestions, and satisfaction levels

- Collect information to identify and analyze issues, problems, and potential solutions

- Benchmark "best" practices

Example

CUSTOMER SATISFACTION SURVEY

Instructions: Please mark the responses below which best reflect your most recent visit. Indicate responses with a ✔ or ✘ in the ☐ provided.

	Excellent				Poor	
	6	5	4	3	2	1

1. How would you rate Video One on an overall basis? ☐ ☐ ☐ ☐ ☐ ☐

2. How would you rate our service package?

	6	5	4	3	2	1
Check-Out	☐	☐	☐	☐	☐	☐
Prices	☐	☐	☐	☐	☐	☐
Staff	☐	☐	☐	☐	☐	☐
Store Location	☐	☐	☐	☐	☐	☐
Supply of New Releases	☐	☐	☐	☐	☐	☐
Title Selection	☐	☐	☐	☐	☐	☐

3. How would you rate our selection of titles?

	6	5	4	3	2	1
Action	☐	☐	☐	☐	☐	☐
Children	☐	☐	☐	☐	☐	☐
Comedy	☐	☐	☐	☐	☐	☐
Documentary	☐	☐	☐	☐	☐	☐
Foreign	☐	☐	☐	☐	☐	☐
Horror	☐	☐	☐	☐	☐	☐
Romance	☐	☐	☐	☐	☐	☐
Sci-Fi	☐	☐	☐	☐	☐	☐

4. Would you recommend Video One to others?

   Yes ☐   No ☐

5. Did you encounter any problems during your last visit?

   Yes ☐   No ☐

   If yes, please explain: _____

   _____

6. What is your age?

   Under 19 ☐   40–49 ☐

   20–29 ☐   50–59 ☐

   30–39 ☐   Over 60 ☐

7. Are you male ☐ or female ☐ ?

FIGURE 8.2 Survey: Video One Customer Satisfaction Survey

Steps

1. Define the objective of the Survey.

2. Identify the types of information and data that are required.

3. Determine the method for conducting the Survey, i.e., verbal (in person or by telephone) or written (by mail, fax, or a combination).

4. Develop the list of questions and arrange in logical sequence. Key considerations in developing a Survey (see Chapter Fourteen: Questionnaires) include:

   • Introduction describes the purpose and scope of the study. It also includes clear, simple instructions.

   • Body of the Survey begins with general, non-threatening questions and ends with more sensitive questions. Avoid words or phrases that might bias responses.

   • Demographic questions conclude the study.

   • Each question should focus on a single specific subject; avoid "nice to know" questions.

   • Break compound sentences into short, simple questions.

   • Avoid jargon and acronyms that are not defined.

   • Group questions by subject and response format within sections.

   • Keep the questions in a logical sequence.

   • Limit branching questions that lead to a specific question or skip a group of questions.

5. Pre-test the Questionnaire to ensure that all questions are necessary, relevant, and not difficult to answer.

Interviewing Tips

Personal Interviews

1. Determine the appropriate contacts by investigating both internal and external sources.

2. Telephone each contact to:

   • Explain the purpose of the study

   • Provide a brief summary of time and resource requirements

   • Identify the expected benefits

   • Arrange a convenient meeting time

3. Before the visit:

   • Send the Questionnaire if pre-work is required before the meeting

   • Prepare responses for your own organization

   • Understand why all questions are being asked

   • Develop back-up questions that might be asked during the interview

   • Plan the visit, including who will participate (at least two people)

   • Identify roles: primary interviewer and primary note-taker

4. During the visit:

   • Adhere to the original time allocation and purpose of the meeting

   • Do not ask questions outside the scope of the study

   • Ask back-up questions and questions that test understanding of answers

   • Take notes

5. After the visit:

   • Debrief interview (assuming two or more participants)

   • Send a thank-you letter to the person interviewed

   • Complete a trip report to include with the study

   • Send final report summary if promised

Telephone Interview

1. Determine the appropriate contacts by investigating both internal and external sources

2. Telephone each contact to:

   • Identify who you are and who referred you

   • Explain the purpose of the study

   • Provide a brief summary of time and resource requirements

   • Identify the expected benefits

   • Arrange a convenient time to conduct the telephone interview

3. Before the telephone call:

   • Send the Questionnaire if pre-work is required before the meeting

- Understand why all questions are being asked
- Develop back-up questions that might be asked during the interview

4. During the telephone call:
   - Adhere to the original time allocation and purpose of the call
   - Do not ask questions outside the scope of the study
   - Ask back-up questions and questions that test understanding of answers
   - Take notes
   - Thank respondent for participation

5. After the telephone call:
   - Complete a telephone call report to include with the study
   - Send final report summary if promised

CHAPTER NINE

ANALYZING CAUSE AND EFFECT

Analyzing cause and effect becomes a pivotal activity in problem solving and process improvement. It links the identified problem or improvement opportunity (i.e., the "as is" or current situation) with its root cause (those elements that drive performance and need changing).

Organization effectiveness depends on two key elements: judgment and execution, that is "doing the right things" and "doing things right." Cause-and-effect analysis not only helps to identify the deficient element, but also help focus attention on the potential causes for which more data might be needed.

Three tools are useful in helping to identify cause-and-effect relationships:

- Cause-and-Effect Diagram (also known as a fishbone chart or Ishikawa diagram)

- Five Whys (using a Tree Diagram)

- Interrelationship Digraph

Cause-and-effect analysis using these tools is often followed by a Matrix Diagram (Chapter Ten) that helps to identify the strength of the causal relationship. When analyzing cause and effect, it is essential that the team have the cross-functional expertise found only in those who are either process owners or process stakeholders. Team members without this expertise should not be excluded, however, because they often add fresh ideas upon which subject matter experts can build.

Throughout cause-and-effect analysis, it is helpful to test causal relationships by asking "Did the causal idea, action, or event precede the effect?" If the answer is no, it is not a cause.

CAUSE-AND-EFFECT DIAGRAM

Description

A Cause-and-Effect Diagram is a graphic technique used to identify and relate possible causes with effects. The effect under analysis can be either the current "as is" state that needs to be corrected or the "desired state" sought. Its strength in analyzing relationships lies in the structured way in which it is developed by using categories of causes that help focus attention beyond symptoms to root, or primary, causes.

The first application of the Cause-and-Effect Diagram to Statistical Process Control was by Dr. Kaoru Ishikawa in 1950. He found that this diagram was helpful to Japanese personnel in organizing the factors that influence a process. In addition to being called an Ishikawa diagram, it is also known as a fishbone chart because of its appearance.

Key Points

- Effective for team problem solving or dissecting complex problems

- Separates and relates problem causes into logical categories

- Applicable to a wide variety of problems

- Helps to focus attention on the vital few root causes

- Helps to separate causes due to either judgment or execution

- Graphically represents the relationships that exist between effects and their causes

- Participants often find it easier to generate causes when the effect is the "as is" statement

Typical Applications

- Identify major and minor reasons or causes for a specific problem or condition

- Identify root causes, or key drivers, contributing to some effect, or measurable outcome (a performance indicator)

- Identify key causes for which additional data are required

Example

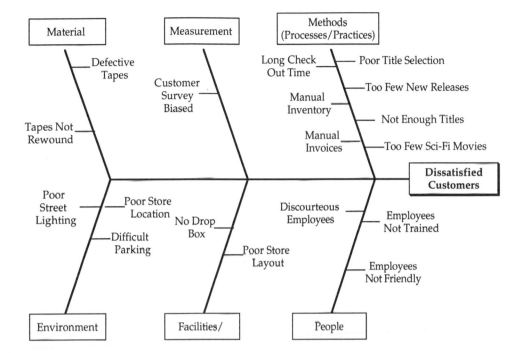

FIGURE 9.1 Cause-and-Effect Diagram: Causes of Dissatisfied Video One Customers

Steps

There are basically two formats: the dispersion format (the standard fishbone chart shown in Figure 9.1) and the process format. The latter format starts with a macro-level Flowchart of key process steps that lead to a problem or undesirable effect. Then, dispersion charts are constructed at each process step. The objective is to identify not only the causes, but also the most likely process step associated with root causes.

1. Clearly state and identify the effect to be analyzed (e.g., an "as is" statement or a "desired state"). Ensure that everyone understands the problem or effect being analyzed. The team must include members with subject matter expertise in the problem or opportunity.

2. Draw the diagram structure (a "fishbone") with the effect in a box at the right side (or "head of the fish") and various categories (or major "bones") branching off the "backbone." Provide sufficient space to diagram factors and their relationships in handwriting that is both large enough to see and

legible. Diagram symmetry, neatness, and the "fish" shape are only secondary considerations.

3. Use the following categories for the headings of the major "bones" or generate more appropriate ones using Brainstorming. Headings should be both logical and inclusive.

   - People
   - Machine (facilities/equipment)
   - Material
   - Methods (processes/practices)
   - Measurements
   - Environment (e.g., culture, organization structure, and technology)

   Service organizations might substitute the following categories:

   - People
   - Processes
   - Policies
   - Facilities/Equipment
   - Material
   - Environment

   Select unbiased categories that do not suggest a cause or a solution.

4. Brainstorm causes, or use other information gathering means, for each of the main "bones." Strive for brief descriptions of causes. Two primary alternatives exist for placing causes on the diagram:

   - Brainstorm a list of causes without concern for the categories, and then place the causes on the diagram
   - Brainstorm causes and add directly to the "bones" identified on the diagram

   (Note that the Affinity Diagram can be substituted for steps 3 and 4.)

5. Add causes beginning with one factor (or "bone") before proceeding to the next "bone." Add causes and subcauses to the diagram by asking, "Why does this happen?" or "What causes this?" Each major "bone," or cause, may have many "sub-bones" and even "sub-sub-bones."

6. Causes related to multiple "bones" should be listed in all appropriate locations.

7. Identify potential root causes. Causes that are repeated on the diagram are often ones that need to be analyzed further.

8. Collect additional data to verify the root causes.

FIVE WHYS

Description

The Five Whys technique is an extension of the cause-and-effect tool above. It can be used as a more structured means of completing step 5 in a Cause-and-Effect Diagram. The structure results from the use of the Tree Diagram, where the effect leads to many branches of the tree, or potential causes, each of which may have many additional branches or causes. The object is to identify root, or key, causes. It is similar in concept to peeling an onion, where each layer of causes leads to additional layers until the core or key causes are defined. (Because it is a similar and supportive tool to cause and effect, only an abbreviated listing is included.)

Key Points

- Effective for team problem solving or dissecting complex problems

- Graphically represents the relationships that exist between effects and their causes

- Provides a more structured, systematic approach to cause-and-effect analysis than the standard Cause-and-Effect Diagram

Typical Applications

- Identify major and minor reasons or causes for a specific problem or condition

- Identify root causes, or key drivers, contributing to some effect, or measurable outcome (a performance indicator)

- Identify key causes for which additional data are required

Example

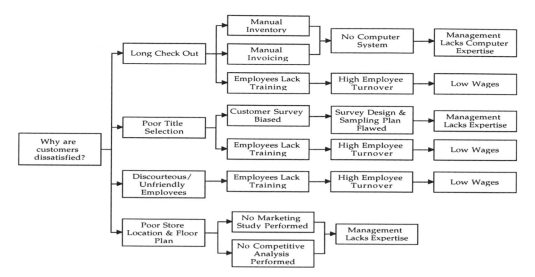

FIGURE 9.2   Five Whys: Causes of Dissatisfied Video One Customers

Steps

See steps for the Tree Diagram and Cause-and-Effect Diagram.

INTERRELATIONSHIP DIGRAPH

Description

The Interrelationship Digraph (or relationship diagram) is a graphic technique that is used to determine the relationships between a given issue or problem and the factors that might cause it. The Interrelationship Digraph shows that most issues are not linear in nature, but instead contain various interrelationships.

When all of these interrelationships are graphed, clusters of relationship arrows indicate root causes. This tool generally requires a high degree of subject matter expertise to be effective and becomes cumbersome when more than 50 items are being evaluated. It is often used in conjunction with the Affinity Diagram and adds logic to the "gut feelings" of the Affinity Diagram.

Key Points

- Illustrates interrelationships between many ideas

- Identifies major factors, or causes, affecting an issue or problem

- Enables problems to be examined from a broad perspective

- Determines in what order issues need to be addressed

- Eliminates preconceptions about major causes of a problem or issue

Typical Applications

- Identify root causes, or key drivers, contributing to some effect, or measurable outcome (a performance indicator)

- Determine both primary and secondary causes of a given effect

- Determine the interrelationship of a multitude of items that have a non-linear relationship to each other

- Identify key causes for which additional data are required

- Identify cause patterns to prioritize problems in critical areas

Example

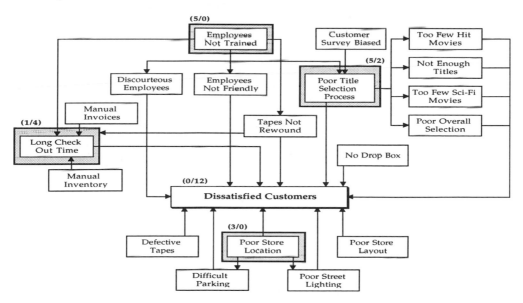

FIGURE 9.3   Interrelationship Digraph: Causes of Dissatisfied Video One Customers

Steps

1. Clearly state and identify the effect to be analyzed (e.g., an "as is" statement or a "desired state"). Ensure that everyone understands the problem or effect being analyzed. The team must include members with subject matter expertise in the problem or opportunity.

2. Write the issue or problem on a 3 × 5 Post-It™ note or index card. Place the statement in the center of a flip chart (use butcher paper or flip chart

paper on a table if using index cards). Limit the number of items being analyzed to less than 50.

3. Generate ideas using Brainstorming or Affinity, Tree, or Cause-and-Effect Diagrams. Use a minimum number of words to express ideas. It is important that several layers of causes be explored before using the Interrelationship Digraph. This prevents the analysis from becoming just multiple causes with only one effect.

   If an Affinity Diagram is used to identify potential causes of a problem, remove the header cards and arrange the remaining items around the problem statement. Ideas that are believed to be most closely related to the problem should be placed closest to it. Alternatively, the cards may be placed randomly on a flip chart.

   In order to avoid confusion, it helps to begin with just a few items, identify the relationships (in step 4), and add items one at a time until all items are included.

4. Using arrows that point in one direction only, draw lines to connect related ideas. Arrows should point *toward* ideas that are the effects and *away* from ideas that are the causes. The problem statement is just another idea in this part of the exercise. The key question to ask is, "Does this idea have a relationship with another item displayed?" Draw two separate arrows, rather than a two-way arrow, between two items that appear to have both a cause and effect relationship on each other.

5. After all the interrelationships have been identified, count the number of arrows pointing toward and pointing away from each note or card. Write these numbers at the top of the card (number away/number toward).

6. The Post-It™ note or index card with the most arrows pointing *away* from it is generally a key causal factor; draw a box around it to highlight this fact.

7. The note or card with the most arrows pointing *toward* it is generally the key effect, or result; draw a box around it to highlight this fact. It often is the real issue and may not be the same one identified in step 1.

8. Identify the next four or five notes or cards with the next highest totals and draw boxes around each one. Use a Criteria Rating Form (Chapter Six) or a Prioritization Matrix (Chapter Ten) to help prioritize the items or issues to be addressed first. Use the number of arrows to help reach consensus as opposed to make the decision.

9. Distribute a legible copy of the digraph to all members for their review.

10. Begin solving the problem or making improvements by focusing effort on the main causal factors identified by the Interrelationship Digraph. This often begins by collecting additional data to verify root causes.

CHAPTER TEN

ANALYZING AND DISPLAYING DATA

The analysis and display of data plays a critical role in the quality improvement process. The proper use of analytical and statistical tools can often mean the difference between success and failure; the pitfalls in using them parallel their importance. Review the section on Validity, Reliability, and Bias in Chapter Three if you have doubts about the pitfalls of data collection and analysis.

Table 10.1 is repeated from Chapter Three and provides a guide to the data types used in several of the tools that follow in this section.

This chapter includes the following analytical tools and techniques:

- Charts (Bar, Pie, Run, and Spider)

- Force Field Analysis

- Histogram

- Matrix Diagram

- Pareto Chart

- Prioritization Matrix

- Scatter Diagram

- Statistical Process Control: Control Charts

- Statistical Process Control: Process Capability

TABLE 10.1   Data Types for Analytical Tools

Tool	Data Types
Bar/Column Chart	Base line: Nominal, ordinal, or yes/no Scale line: Discrete interval
Pie Chart	Nominal labels Slices: Continuous interval percentages
Run Chart	Base line: Interval or ordinal Scale line: Interval
Histogram	Base line: Interval Scale line: Discrete interval
Pareto Chart	Base line: Nominal or ordinal Scale line: Discrete interval
Scatter Diagram	Base line: Interval Scale line: Interval
Control Charts	Base line: Ordinal or discrete interval Scale line: Continuous interval (Variable Charts) and discrete interval (Attribute Charts)
Process Capability	Base line: Continuous interval Scale line: Discrete interval

CHARTS (BAR, PIE, RUN, AND SPIDER)

Description

Graphic displays provide insight that often is not possible with words or numbers. Six of the seven original quality tools are graphic techniques for analyzing and displaying data. Two of the tools are specialized forms of vertical Bar (or column) Charts: Histograms display frequency of interval data and Pareto Charts show ordinal, nominal, and discrete interval data. Control Charts are a specialized form of a Run Chart, or time line chart, where there are statistically determined control limits above and below the average measurement.

Bar Charts display comparisons of data categories (nominal, ordinal, or yes/no data on the base axis compared with discrete interval data on the scale axis). Bar Charts display bars running either vertically or horizontally and have many formats, including stacked data, multiple bars, and combinations with line graphs. The bars can represent categories at a point in time or changes in categories over a period of time (shown as multiple columns).

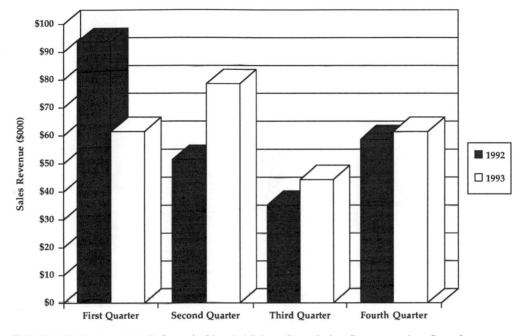

FIGURE 10.1   Bar (or Column) Chart: Video One Sales Revenue by Quarter

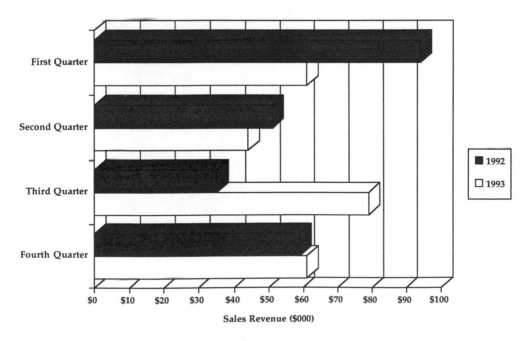

FIGURE 10.2   Bar Chart: Video One Sales Revenue by Quarter

Pie Charts are circle graphs that display 100% of the data as a circle. The circle is divided into slices that represent nominal categories whose size is defined by the category percentage of the total.

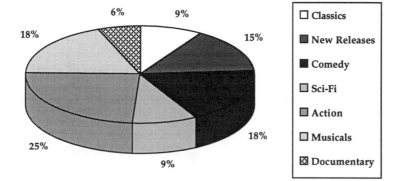

FIGURE 10.3   Pie Chart: Video One Rental Categories

Spider Charts, or radar charts, show values for various types of nominal data where each nominal category is compared using the same scale. Each category has its own axis radiating from the center point. When data points belonging to the same series are connected, the chart looks like a spider web. The axis has either a continuous or discrete interval scale.

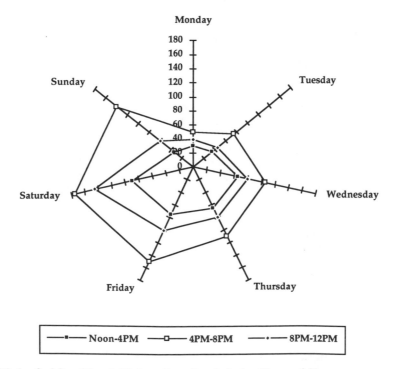

FIGURE 10.4   Spider Chart: Video One Rentals by Time of Day

Run Charts

Run Charts (or time line charts) show trend data for a process over time. The horizontal axis (x) represents ordinal or interval data (i.e., the time line) and the vertical axis (y) has an interval scale, either discrete or continuous. The data displayed may be a single line or multiple lines reflecting the various characteristics, or strata, of the data. Breaking the data into separate strata illustrates the differences between each element and the total. The balance of this section refers to Run Charts.

Key Points

- Simple to construct and use
- Provides a visual perspective of a process variable over time
- Often combined with other graphs to visually relate change in common or related process variables
- Does not indicate whether a process is in control or out of control
- Provides very limited capability for determining whether variation is due to special or common causes

Typical Applications

- Visually track process variation over time
- Identify trends, shifts in the average, or other process variables, either separately or in combination with other graphic displays, such as Control Charts and Histograms

Example

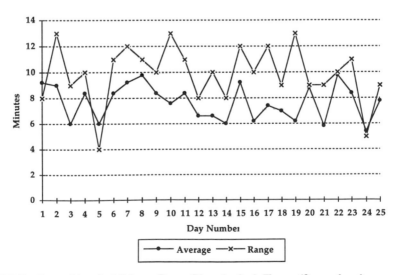

FIGURE 10.5 Run Chart: Video One Check-Out Time (Sample Averages and Range)

Steps

1. Use a Check Sheet to collect data on a process variable for a period of time. A minimum of twelve data points are necessary to provide a meaningful plot. Although it is desirable to have data points evenly dispersed over time, it is not required because the data will be plotted on a time line of uniform intervals. Keep all data points in sequential order.

2. Mark the vertical axis (y) with the name of the variable measured and its unit of measure. This axis shows interval data, either discrete or continuous.

3. Mark the horizontal axis (x) with the word "Time" or "Sequence" and the appropriate unit of measure, which can be either ordinal or time-related interval data.

4. Label the chart with meaningful title(s) to describe what is measured and during what time frame.

5. Divide and label the horizontal and vertical axes into evenly spaced increments.

6. Plot and connect the data points. New data points can be added as they become available.

7. Add grid lines, data point labels, and a data centerline (mean) if they improve clarity or aid in understanding the data. If grid lines are added, use them only on major increments to avoid a cluttered look that would detract from the visual image of the plotted line.

8. Do not use Run Charts in lieu of Control Charts, which require a specific Sampling procedure (e.g., timing, subgroup size, and Sampling methodology). If the Run Chart uses the same procedures, it can become a Control Chart.

Interpretation

Do not be alarmed by variations associated with the natural variation of the process. Only some of the variation shown on a Run Chart is due to unusual circumstances (special causes) and needs attention. Run Charts are not well suited for discovering these special causes, however. The distinction between common and special causes can only be determined when sufficient data are available to construct a Control Chart.

Run Charts of sample averages can be interpreted for some fairly simple statistical tests:

1. Run Charts of sample averages should have an equal number of points above and below the data centerline (the process average). Whenever nine or more points in a *row* remain ("run") on one side of the data centerline, this indicates that a process change has occurred.

2. A trend of six or more points *continuously* increasing (or decreasing) also indicates that the process is changing. The first point that starts the trend is not counted.

3. Fourteen consecutive points alternating up and down on a chart is statistically improbable. The first point that starts the trend is not counted.

FORCE FIELD ANALYSIS

Description

Force Field Analysis is a technique used to identify forces that either help or hinder change. It is a creative activity to help teams focus on change from the current state (i.e., "as is" condition) to the "desired state." This technique highlights both weaknesses, or barriers to change, as well as strengths that aid change. Furthermore, it provides a subjective assessment of the strength of opposing forces.

Although Force Field Analysis can be used to evaluate problems or improvement opportunities in Step 1 of the Quality Improvement Model, it is more commonly used to evaluate forces that affect alternate solutions in Step 4.

Key Points

- Uses Brainstorming to identify forces that help and hinder change
- Identifies helping forces or conditions that need to be exploited
- Identifies restraining forces, or barriers, that cause problems
- Helps team choose between opposing forces (i.e., to remove hindering forces or strengthen helping forces)
- Helps team focus on the relative importance of the opposing forces
- Often used as a strategic tool for change

Typical Applications

- Determine opportunities and barriers to quality improvement
- Identify organization strengths and weaknesses
- Identify problems as well as opportunities to be exploited
- Assist in the analysis of cause and effect
- Help analyze and prioritize solutions to problems or areas needing improvement

Example

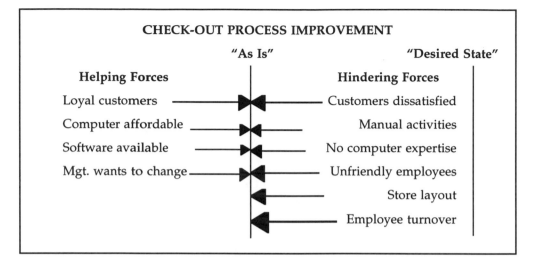

FIGURE 10.6   Force Field Analysis: Video One Check-Out Process Improvement

Steps

1. State the topic of the Force Field Analysis; clearly write the topic at the top of a flip chart or board.

2. Draw two vertical lines, one in the center of the chart and the other on the right side. Label the center one "As Is" and the right one "Desired State." Underneath these labels, write "Helping Forces" on the left and "Hindering Forces" on the right.

3. Explain helping and hindering forces. Those that aid the change process are helping forces and those that are barriers hinder change.

4. Brainstorm forces for the selected topic. Record all forces on the flip chart or board in large enough text that everyone can see. List each force under its respective category (helping on the left side and hindering on the right side). Use List Reduction to reduce the list to the most important forces (usually ten or fewer).

5. Discuss each force and reach consensus on the relative strength of forces. Use a simple 1 to 5 scale (where 5 is strongest), and rate all forces (i.e., rate the magnitude of the force).

6. Draw arrows pointing to the left for hindering forces and to the right for helping forces. The length of the arrow varies with the strength of the force. Arrowheads end at the centerline. Some forces may appear to both help and hinder. This often indicates that a force needs to be analyzed in greater detail to separate the conflicting elements.

7. The team can now focus on strengthening helping forces or on removing hindering forces.

HISTOGRAM

Description

A Histogram is a chart that displays the frequency distribution of one measure or characteristic of data from a process. Unlike the Pareto Chart, which generally displays nominal or ordinal data, the Histogram is used to display the patterns of variation of either discrete or continuous interval data. These pictures of data illustrate patterns of variation, which aid in understanding a process.

Control Charts (see Statistical Process Control later in this chapter) are constructed, in essence, from a series of Histograms, one for each set of data or subgroup of data. After a certain number of data sets (usually 25) are collected, a Control Chart can be developed. Because Histograms provide additional information about a process often not evident from Control Charts or Run Charts, they are used as a corollary tool for process analysis.

Key Points

*   Provides an easy-to-understand means of displaying the variability of data

*   Does not show change in data over time, but represents a snapshot at a certain point in time; use Run or Control Charts for trend analysis

*   Histogram shape often provides information that is not evident from Control Charts

*   Visual interpretations of Histograms include the spread (i.e., variability), skew (i.e., normal or skewed, either right or left), and uniformity of shape

Typical Applications

*   Display and compare process variability with expected variability

*   Verify whether a process is normally distributed or skewed

*   Determine whether two machines, processes, etc. are producing with the same median, mean, and variability

*   Provide visual information that helps to interpret the output of a process and to understand both common and special causes of variation when used with Control Charts (see Interpretation below)

Example

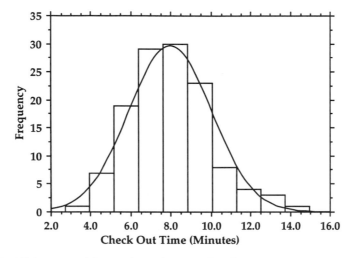

FIGURE 10.7   Histogram: Video One Check-Out Time

Statistical Definitions

Mean

The mean, or average, is the sum of all values counted divided by the number of items counted. It is commonly designated as x-bar ($\overline{X}$).

Median

The median is the value halfway between the highest and the lowest data values when all data values are listed in ascending or descending order. With an even number of data points, the median is defined as the average of the two midpoint values.

Mode

The mode is the most frequently occurring value of the data set.

Normal Distribution

A normal distribution is a symmetrical, bell-shaped distribution in which the mean, median, and mode are all the same.

Furthermore, if the total area under the curve is assigned a value of one, then the area under any portion of the curve can be calculated and converted to probabilities. A normal curve is illustrated in Figure 10.8, along with a Histogram generated from samples taken from a normally distributed population.

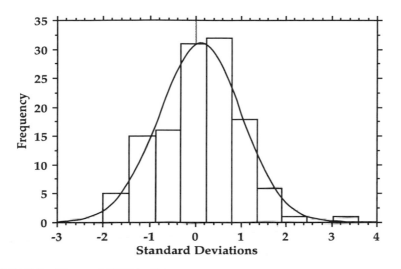

FIGURE 10.8   Normal Distribution

Range

Range is a measure of the variability that exists in a data set. It is equal to the lowest value in a data set subtracted from the highest value.

Standard Deviation

Standard deviation is a measure of the variability that exists in a data set and is commonly represented by the lowercase Greek letter sigma (σ). In a normal distribution, 68.26% of the data will be within one σ from the mean, 95.45% within two σ, and 99.73% within three σ. Mathematically, a standard deviation is equal to the square root of the average squared differences between individual data values and the data set average.

$$\sigma = \sqrt{\frac{(X_1 - \overline{X})^2 + \dots + (X_n - \overline{X})^2}{(n - 1)}}$$

Steps

1. Count the number of data points (N) in the data set. There should be at least 50 data points.

2. Determine the range (R) of the data set by subtracting the lowest data value from the highest.

3. Based on the number of data points, divide the range into a number of equal-size classes using Table 10.2.

TABLE 10.2 Histogram Data Classes

Number of Data Points (N)	Number of Classes (K)
Less than 50	5–7
50–100	6–10
100–250	7–12
Over 250	10–20

4. Calculate the classification width (H) by dividing the range (R) calculated in step 2 by the number of classes (K) chosen from Table 10.2. Round the width to a number with the same number of decimal places as the data. For example, if we have 100 data points carried to one decimal place with a range of 79.3 and we pick 8 classes from the table, then the width will be 79.3 ÷ 8 = 9.9. For simplicity, round up the number to 10.0.

$$H = R \div K$$

5. Next, boundary limits for each class must be determined. In order to ensure that all data points fit into a class, the boundary limits will be carried to one decimal place more than the data. For example, if the data is carried to tenths (0.1), then the boundary limits will be carried to one hundredths (0.01). In this case, 0.01 is a more "refined" number than 0.1 because it is carried to one more decimal place.

To begin the first class boundary and to ensure that the lowest number is included, select the lowest number in the data set and subtract the smallest, more refined number from this value. This becomes the lower boundary of the first class.

For example, if the smallest number is 3.5, then the smallest, more refined number is 0.01 and the lower boundary becomes 3.49. Now add the class width from step 4 to define the boundaries of the first class. In our example, the lowest number is 3.5 and the highest number is 82.8, with a range of 79.3. The first class then becomes 3.49 to 13.49, that is 3.5 − 0.01 = 3.49 for the lower boundary and 3.49 + 10.0 = 13.49 for the next boundary. The second class boundaries are 13.49 and 23.49. The last class boundaries are 73.49 and 83.49.

6. Tabulate a frequency table from the raw data to simplify Histogram construction. Begin the table by listing the class boundaries. Next, place a check, or "tick" mark, next to each class interval to correspond with the

number of data points that fit each class. This is a Histogram in tabular form, with each data point represented by a tick mark in the appropriate class.

7. Finally, construct the Histogram. Label the vertical axis (y) to denote the frequency of the event observed or measured. Identify the class intervals on the horizontal axis (x) and add an appropriate title to describe the data represented. Draw columns to heights that reflect the frequency of occurrences in each class, with no spaces between the columns.

Interpretation

Ideally, the Histogram will create a picture that shows the variation of data values and simultaneously provide a reasonable level of detail. If too few classifications have been chosen, little of the shape (except perhaps for a few large bars) will be seen. Conversely, if too many classes are shown, the shape of the data will be lost since some of the classes will be empty, resulting in a comb-like toothed appearance.

The shape of the Histogram is often the first indication of a problem with a process that needs further investigation. Because these problems may not be identified by a Control Chart, it is important to use Histograms in support of Statistical Process Control. Various Histogram shapes and their interpretations follow.

1. If the Histogram appears to have two peaks (i.e., bimodal), then the data may in fact be coming from two or more *different* sources. In some cases, the peaks may be the same height and in other cases there may be one large peak and a smaller, isolated peak. Another form of double peaking occurs when there is a normal peak and an "edge" peak of almost equal height near the tail of the distribution. An attempt to stratify the data will help to uncover the multiple sources.

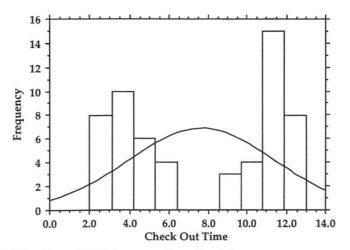

FIGURE 10.9 Bimodal Histogram (Two Peaks)

2. When classes stop abruptly (i.e., truncated shape) without first declining, this is a warning that the data may be in error or that an inspection and/or selection activity is affecting the results. This truncation may occur on both sides of the Histogram. An example of this is sometimes found near reject or specification limits. If the Histogram shows many values immediately below the reject level but mysteriously none at all above the reject limit, then there is a good probability that either the data are being falsified to avoid rejects or that out-of-spec output is being selected out. In either case, the customer is paying for the costs associated with the rejects.

This truncated effect may also show up with a slot out of the center of the Histogram where data points are missing for output that has been selected out. This is done to satisfy a customer with more stringent specifications than the customer preparing the Histogram, who is receiving output within his or her specification but of lesser "quality" than the customer receiving the output from the central slot.

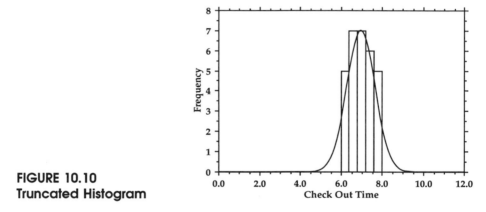

FIGURE 10.10
Truncated Histogram

3. Random data samples for most processes that are in statistical control will follow the bell-shaped normal curve. In this curve, the greatest frequency will be at some center measurement value, with an approximately equal number of points falling on either side of this value. Some processes, however, are naturally skewed. If a normal curve is expected, then examine the conditions in the long tail away from the skewed peak.

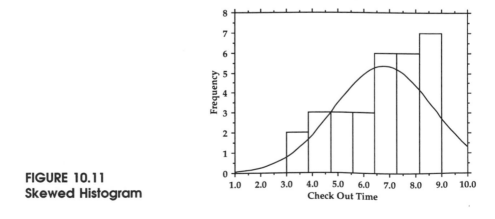

FIGURE 10.11
Skewed Histogram

4. Although the comb-like appearance is often due to Histogram construction problems, there may also be data recording or measurement problems that result in the absence of data points in alternating classes. For example, the measurement equipment may not be able to measure to the level of detail for the classes chosen.

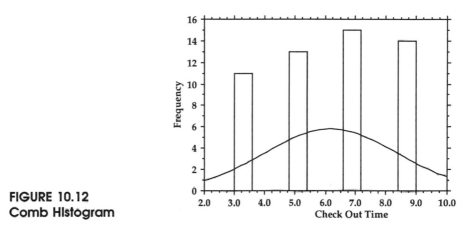

FIGURE 10.12
Comb Histogram

5. A Histogram that is broad and flat (i.e., a plateau) instead of peaked might indicate an ill-defined Histogram, but it may also mean that there are multiple processes being measured.

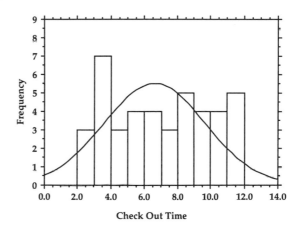

FIGURE 10.13
Plateau Histogram

MATRIX DIAGRAM

Description

The Matrix Diagram is a management and planning tool that is used to analyze relationships between two or more variables. Many of these relationships are between independent and dependent variables, e.g., cause and effect or what and how (means). The main advantage of the Matrix Diagram is its focus on graphically illustrating the strength of relationships between each paired combination of variables. The highly visible symbols used to show the strength of relationships result in easily identifiable visual patterns.

Matrix Diagrams, like other management tools, have multiple uses and are often used with other management tools. The flexibility in use includes applications supporting cause-and-effect analysis, prioritization of alternatives, and planning or deployment of tasks. The Matrix Diagram is particularly helpful in evaluating the strength of cause-and-effect relationships. The detailed elements of Matrix Diagrams often come from previous Affinity and Tree Diagrams.

Prioritization Matrices are extensions of the basic Matrix Diagram.

Key Points

- Matrix Diagrams are commonly denoted by their shapes, which fit into five common, two-dimensional formats: L, T, X, Y, and roof formats. The C matrix, a three-dimensional array, is a sixth format that is used less frequently because of its complexity. The roof format is used to compare the relationships between the elements of one of the two or more variables being analyzed. These relationships are either neutral, positive (complementary), or negative (opposing).

TABLE 10.3 Matrix Diagram Formats

Matrix Format	Number of Variables	Direct Relationships	Indirect Relationships
L	2	1	0
T	3	1	1
Y	3	3	0
X	4	4	2
C	3	3 simultaneous	0
Roof	1	—	—

- Special symbols are used to indicate the strength and importance of a relationship or to identify responsibility and degree of involvement of individuals or organizations with a task. These symbols are sometimes followed by a "+" or "–" to denote whether the relationship is supportive or negative. Arrows are often added to indicate the cause-and-effect relationship of one variable on another. Arrows point toward the element affected. Two-headed arrows should be avoided.

- The graphic format of the matrix and the highly visible symbols used to illustrate relationships create visual patterns that facilitate analysis.

TABLE 10.4 Matrix Diagram Relationship Symbols

Symbol	Symbol Value	Relationship	Supportive (+)/ Negative (−)	Responsibility
⊙	9	Strong	Strong	Primary
○	3	Some	Some	Secondary
△	1	Weak/Possible	Weak/Possible	Keep Informed/ Consult for Input

Typical Applications

- Analyze relationships between variables, such as causes and effects
- Prioritize and select activities, functions, projects, and processes for improvement
- Allocate resources for high-priority activities, functions, projects, etc.
- Deploy definable and assignable tasks (how) to specific individuals or organizations (who) in order to achieve objectives and goals (what) using a Y Matrix Diagram
- Use to link customer needs and expectations (what) with various characteristics and functions (how)
- Analyze conflicts and explore substitutes among opposing functions or characteristics

Example

L Shaped Matrix   **Roof Shaped**   **Y Shaped**

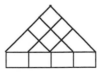

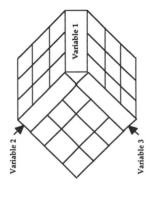

T Shaped Matrix   **X Shaped Matrix**

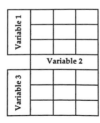

FIGURE 10.14 Matrix Diagram Shapes

PROCESSES

CUSTOMER NEEDS & EXPECTATIONS	Survey Design & Analysis	Check Out	Title Selection & Ordering	Advertising	Competitive Intelligence	Information Systems	Employee Training	TOTAL
Low Prices		△	O	O	◉	O	O	22
Good Title Selection	◉		◉	O	△	O	◉	34
Multiple Copies	◉		◉			O		21
Good Store Layout		◉			O			6
Good Store Location					O			3
Fast Check Out		◉				◉	O	21
Friendly Staff		O					◉	12
TOTAL	18	22	21	6	16	18	24	

Strong	◉	9
Some	O	3
Weak	△	1

FIGURE 10.15 L Matrix Diagram: Video One Matrix of Customer Expectations versus Work Processes

Steps

1. Identify the variables for which relationships must be established (e.g., specific tasks and individuals, customer needs and expectations, processes, functions, characteristics, etc.). Matrix Diagrams usually require subject matter expertise in selecting the variables and in evaluating relationships. Assemble the right team, including individuals who are process owners or stakeholders.

2. Select the matrix format based on the number of relationships, both direct and indirect, using Table 10.3.

3. Identify the variables and their elements as the headers of columns and rows in the matrix. The elements of the variables might be from earlier Affinity Diagrams or Tree Diagrams or from a new Brainstorming activity.

4. Complete the matrix format by drawing in the grid lines. If the Matrix Diagram is to be completed as a result of a Tree Diagram, simply draw a grid to the right of the last level of detail on the Tree Diagram.

5. Determine the symbols (including supplemental arrows and "+" or "−" signs) to be used to denote relationship between the two variables. Use symbols from Table 10.4 rather than numbers to allow visual patterns to emerge. Do not use two-way arrows. Instead, point the arrow toward the item most strongly influenced.

6. Fill in the matrix by entering the appropriate symbol in each cell where a relationship exists. Convert symbols to their defined values and sum values for each row and column. Show row and column totals.

7. Review the completed matrix. If the matrix is being used to assign tasks in a plan implementation, look for balance in tasking, areas of responsibility not assigned, and unclear divisions of responsibility.

8. Ensure that the relationships depicted between the variables are understood and agreed to by all team members.

PARETO CHART

Description

A Pareto Chart is a vertical Bar Chart that displays the relative frequency of various categories of a problem or condition. Because the bars are arrayed in descending magnitude, from left to right, the Pareto Chart helps to identify the "vital few" items that contribute to, or cause, the majority of the problem or condition measured. The Pareto Chart is usually accompanied by a cumulative percentage line that depicts the cumulative frequency of items, starting with the most frequent category and adding remaining categories until 100% is reached.

Often, 20% of the items contribute to 80% of the effect; hence, the "80–20" rule. The focus of problem solving and process improvement is generally on the vital few, because correcting or changing them will often have the largest payback. This search for inequalities in causal factors is based on the work of the 19th century economist Vilfredo Pareto.

Pareto Charts require the collection of data using Check Sheets or other data collection forms. In some cases, the data categories need to be varied or stratified in order to uncover the 80–20 relationship. This stratification often requires that new categories be formed in order to separate the data on the basis of location (physical and geographical), time, key characteristics (e.g., dimensions, degree, and duration), or individuals and organization.

Key Points

*   Based on the Pareto principle, or "80–20" rule, which states that a few causes are responsible for most of the effect

*   Provides rank-ordered list of causes and their relative contribution, a "snapshot" of a process and its problems

*   A 50% reduction of the most common cause is usually easier to accomplish and more beneficial than eliminating the least common cause

*   Choosing the right categories is critical to creating Pareto Charts. For example, a bottleneck activity will limit all other activities in a process. A Pareto Chart that focuses on actual cost, versus lost opportunity cost, will not highlight the bottleneck if its costs are low. (Lost opportunity cost of a specific activity is defined here as the highest capacity activity in the

process minus the capacity of the specific activity times the unit profit of the process.)

Typical Applications

- Display causes of problems in order of importance
- Verify root causes of problems after cause-and-effect analysis
- Display "before" and "after" data for process improvement projects

Example

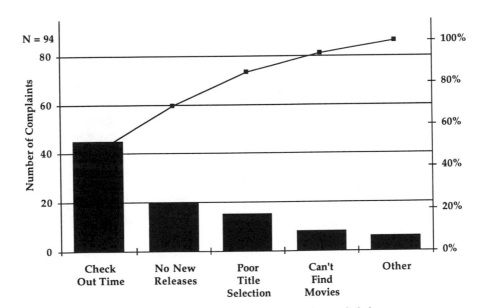

FIGURE 10.16   Pareto Chart: Video One Customer Complaints

Steps

1. Select the problem or process to be studied. This often occurs after cause-and-effect analysis in Step 3 of the Quality Improvement Model. Determine the types of categories or classifications into which the data will be divided.

2. Over a specified time period, collect data that define either the incidence or quantity of each category, such as the number of failures, complaints, or costs. Subtotal the data and rank in descending order. For each category, calculate its percentage of the total data set (e.g., total items counted or total costs).

3. Mark the left vertical axis (y) with the name and unit of measure of the incident being measured. Divide this axis into equal intervals using a scale ranging from zero to the sum of all occurrences recorded. The number of customer complaints shown in Figure 10.16 totals 94, which also corresponds with the 100% cumulative percentage line (see step 7 below).

4. Divide the horizontal axis (x) axis into equally sized intervals, one for each category identified. Generally, no more than ten bars should be plotted on this axis. If there are more than ten categories, then the remaining categories should either be omitted or accumulated into an "Other" category.

5. Plot the data for each category as bars, with the most frequent category at the far left and continuing sequentially down to the smallest category (or "Other") on the far right. Identify each bar with a descriptive label for easy identification. "Other" can be more frequent than the second-to-last item because it is an accumulation of remaining categories.

6. Label the chart with a meaningful title and any other pertinent information.

7. Draw and mark the right vertical axis to show the percent of total, with the scale running from 0% to 100%. Plot the cumulative frequency (i.e., cumulative percentage of total). Start with the top of the tallest bar and add percentages for each successive category to the right until the 100% point is reached. The 100% point on the right axis corresponds with the sum of all occurrences on the left scale.

8. If the Pareto Chart does not separate the vital few from the trivial many, then try a different stratification of the data. This stratification often requires that new categories be formed to separate data on the basis of location (physical and geographical), time, key characteristics (e.g., dimensions, degree, and duration), or responsibility.

PRIORITIZATION MATRIX

Description

The Prioritization Matrix is a management and planning tool that is used to prioritize and select tasks, projects, and issues based on weighted criteria of importance. Although related in concept to the Criteria Rating Form (Chapter Six: Consensus), this tool is capable of handling more complex, detailed analyses. It is therefore less likely to be used by work unit teams to reach consensus than it is to be used by a management team trying to prioritize actions for an organization larger than a single work unit.

The use of a Prioritization Matrix generally follows four basic steps: selection of alternatives, selection and ranking of criteria, ranking of alternatives on the basis of each criterion (i.e., one at a time), and ranking of all alternatives by all criteria. Alternatives often come from the lowest level of a Tree Diagram. Criteria for rating alternatives are similar to those discussed under Criteria Rating Forms (Chapter Six). Ranking of criteria and ranking of alternatives by criterion use an L-shaped Matrix Diagram in a process that is similar to, but more analytical than, the Paired Comparison tool (Chapter Six).

The Prioritization Matrix selects alternatives based on some criteria but does not measure how effectively the criteria are satisfied and what level of improvement is required. Organizations using the Prioritization Matrix to select processes to improve need to analyze current performance (i.e., effectiveness) and the level of improvement required before making a final decision.

The Prioritization Matrix replaces the Matrix Data Analysis as one of the seven management tools. Although there are several formats of Prioritization Matrices, only two will be presented here: the Analytical Criteria Method (similar to the Analytical Hierarchy Process) and the Consensus Criteria Method. A Combination Interrelationship Digraph/Matrix Method (presented in *Memory Jogger Plus+™*) is not included.

Key Points

- Compares alternatives using the same criteria

- Generally applied to more complex decisions (i.e., more criteria and alternatives) than the Criteria Rating Form

- Uses criteria that are often less controversial, and therefore easier to reach consensus on, than the alternatives themselves

- Enables the weighting of important criteria

- Provides a logical basis for reaching consensus necessary for a decision

Typical Applications

- Identify and select high-priority tasks, projects, and issues based on weighted selection criteria

- Rank closely related alternatives

- Resolve disagreements over decision criteria

- Focus on critical problems and processes given limited resources

Example: Analytical Criteria

	CRITERIA						
CRITERIA	Control	Importance	Difficulty	Resources	Payback	TOTAL	PERCENT
Control		1/10	1	1	1/5	2.3	4.0%
Importance	10		10	10	5	35.0	61.3%
Difficulty	1	1/10		1	1/5	2.3	4.0%
Resources	1	1/10	1		1/5	2.3	4.0%
Payback	5	1/5	5	5		15.2	26.6%
TOTAL	17.0	0.5	17.0	17.0	5.6	57.1	100.0%

> 1 = Equal Importance
> 5 = Significantly More Important
> 10 = Extremely More Important
> 1/5 = Significantly Less Important
> 1/10 = Extremely Less Important

NOTE: Only Importance and Payback are significant for ranking alternatives.

FIGURE 10.17   Prioritization Matrix: Ranking of Video One Decision Criteria

	ALTERNATIVES					
ALTERNATIVES	Survey Design & Analysis Process	Check Out Process	Title Selection & Ordering Process	Advertising Process	TOTAL	PERCENT
Survey Design		1/10	1/5	1	1.3	3.8%
Check Out	10		1	5	16.0	46.2%
Title Selection & Ordering	5	1		10	16.0	46.2%
Advertising	1	1/5	1/10		1.3	3.8%
TOTAL	16.0	1.3	1.3	16.0	34.6	100.0%

FIGURE 10.18   Prioritization Matrix: Ranking of Video One Alternatives by Criterion ("Importance")

SIGNIFICANT CRITERIA *

ALTERNATIVES	Importance	Payback	TOTAL	PERCENT
Survey Design & Analysis	0.038x0.613	0.132x0.266	0.058	6.6%
Check Out	0.462x0.613	0.487x0.266	0.413	47.0%
Title Selection & Ordering	0.462x0.613	0.371x0.266	0.382	43.4%
Advertising	0.038x0.613	0.010x0.266	0.026	3.0%
			0.879	100.0%

* Ranking by significant criteria only

FIGURE 10.19   Prioritization Matrix: Ranking of Video One Alternatives by All Criteria

Steps

1. Identify and select the alternatives to be prioritized using idea generation/consensus tools. Tree Diagrams are often used to ensure that alternatives are both organized and at a sufficient level of detail. Clarify all alternatives. Team members must have subject matter expertise in the alternatives being evaluated.

   Next, identify and select criteria to be used in prioritizing the alternatives. Here too, idea generation/consensus tools are used along with standard criteria used for similar evaluations.

Criteria Ranking

2. Create an L-shaped Matrix Diagram (Figure 10.17) with criteria labels on both axes of the matrix. List the criteria in order starting from the upper left corner. The diagonal where each criterion intersects with itself is blacked out.

3. Establish a prioritization scale for the criteria. Typically, 10 is assigned as extremely more important, 5 as significantly more important, 1 as equally important, 1/5 as significantly less important, and 1/10 as extremely less important. Note that the less important scale numbers are the inverse of the more important scale numbers.

4. Rate each row criterion relative to each column criteria; complete the matrix row by row. For consistency, if item #1 relative to item #2 is assigned a 5, then item #2 relative to item #1 must be assigned 1/5 (i.e., inverse or mirror image). Expressing fractions as decimal equivalents often eases the analysis. Team consensus must be obtained for each comparison.

In Figure 10.17, for example, the 1/10 in the Control row and Importance column indicates that the Control criterion is extremely less important than the Importance criterion. Thus, importance is given a value of 10 relative to control.

5. Total the rating assignments across the rows and down the columns. Verify that the grand total for all rows equals the grand total for all columns.

6. Divide each row total by the grand total and record the quotient as a percentage. This percentage is the weighted priority for each criterion.

Ranking of Alternatives by Criterion

7. Create L-shaped Matrix Diagrams (Figure 10.18) for *each* criterion selected. (Not all criteria must be evaluated, because less important ones often do not influence the prioritization and can be ignored. This is often determined in step 6. In our example, only Importance and Payback would be selected based on their relative ranking.) Identify each matrix with the criterion being evaluated and list the alternatives on both axes of the matrix. Again, start in the upper left corner and black out the intersection of each alternative with itself.

8. Establish a prioritization scale for the alternatives in terms of the criterion being evaluated. If Importance is the criterion, as in Figure 10.18, then 10 is assigned as extremely more important , 5 as significantly more important, 1 as equally important, 1/5 as significantly less important, and 1/10 as extremely less important. Note that the less important scale numbers are inverses of the more important scale numbers.

9. Rate each row alternative relative to each column alternative; complete the matrix row by row as in step 4 above.

10. Total the rating assignments across the rows and down the columns. Verify that the grand total for all rows equals the grand total for all columns.

11. Divide each row total by the grand total and record the quotient as a percentage. This percentage is the weighted priority for each alternative for the criterion evaluated.

Ranking of Alternatives by All Criteria

12. Create an L-shaped Matrix Diagram (Figure 10.19). Here, the alternatives are listed on the vertical axis and the criteria on the horizontal axis. Use only the criteria selected in step 7.

13. Multiply each criterion weight (i.e., percentage from step 6) by the alternative weighting for that criterion (i.e., percentage from step 11). Each box in the matrix will have a value that is the product of two percentages.

14. Total the values across the rows.

15. Divide each row total by the grand total and record the quotient as a percentage. This percentage is the weighted priority for each alternative using all criteria evaluated. In Figure 10.19, Check-Out and Title Selection Processes would be the two alternatives considered for the consensus decision.

Example: Consensus Criteria

ALTERNATIVES	SIGNIFICANT CRITERIA *			
	Importance	Payback	TOTAL	PERCENT
Survey Design & Analysis	1x0.613	2x0.266	1.145	13.0%
Check Out	4x0.613	4x0.266	3.516	40.0%
Title Selection & Ordering	3x0.613	3x0.266	2.637	30.0%
Advertising	2x0.613	1x0.266	1.492	17.0%
			8.790	100.0%

* Ranking by significant criteria only

FIGURE 10.20   Prioritization Matrix: Ranking of Video One Alternatives by All Criteria Using Consensus Criteria Method

Steps

1. Identify and select the alternatives and criteria for evaluation as in step 1 of the Analytical Criteria method above.

Criteria Ranking

2. Use steps 2 through 6 above to establish criteria weighting. The team can limit the criteria selected based on relative importance, as done earlier.

Ranking of Alternatives by Criterion

3. Use Paired Comparisons (Chapter Six: Consensus) to rank all alternatives relative to *each* criterion selected (not all criteria need to be selected, as noted above). The most important alternatives will have the highest values and the lowest ranking will be 1.

    Another approach is for each team member to rank alternatives (highest being the most important) for each criterion and use a weighted average ranking. Team consensus must be obtained for each comparison, however. Again, alternatives are ranked on a criterion-by-criterion basis.

Ranking of Alternatives by All Criteria

4. Create an L-shaped Matrix Diagram (Figure 10.20). List the alternatives on the vertical axis and the criteria on the horizontal axis, as in the Analytical Criteria method. Use only the criteria selected in step 2 (i.e., of the Consensus Criteria method).

5. Multiply each criterion weight from step 2 by the alternative ranking from step 3. Each box in the matrix will have a value that is the product of these two numbers.

6. Total the values across the rows. Alternatives are prioritized by highest row values. In Figure 10.20, the Check-Out and Title Selection Processes would be the two alternatives considered for the consensus decision.

SCATTER DIAGRAM

Description

A Scatter Diagram provides a graphic plot of two variables. The data are generally interval data. The resulting pattern, which can range from a shotgun pattern to a near perfect line of data points, indicates the degree of correlation, or strength of the relationship, between the two variables. The existence of the relationship does not infer a cause-and-effect relationship, however, since a third variable may be the causal factor. The relationships can be either positive (i.e., both variables increase together) or negative (i.e., one increases while the other decreases); they can also be either linear or non-linear.

Correlation determines the strength of a relationship between two variables, while regression analysis is required in order to fit a line or a curve to the data points (i.e., equation of a line or curve determined by the mathematical relationship between variables). Correlation analysis is beyond the scope of this book.

Key Points

- Requires a large number of data points

- Indicates whether or not two process variables are related to one another

- Provides an indication of whether a cause-and-effect relationship might exist

- Typically requires additional investigation (correlation tests and regression analysis) before acting on the data

- Improperly scaled diagrams can provide misleading conclusions

- Expanded data ranges and stratification of data points can often illustrate relationships that might not otherwise be visible

Typical Applications

- Plot suspected causes versus undesirable effects in support of cause-and-effect analysis

- Compare a series of paired variables to determine if a relationship exists

- Indicate whether additional analysis is warranted to determine the exact nature of a cause-and-effect relationship (i.e., to predict the relationship by fitting a line or curve to the data points)

Example

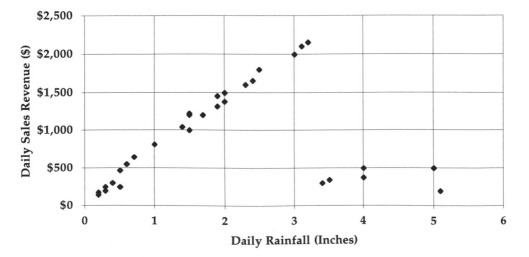

FIGURE 10.21 Scatter Diagram: Video One Daily Sales Revenue versus Daily Rainfall

Steps

1. Obtain data on various process variables. Select two variables suspected of being related (i.e., a dependent and an independent variable). For each value of the independent variable, determine the value of the suspected dependent variable. These two values create a data pair that will be plotted. Typically, Scatter Diagrams require at least 30 but preferably over 100 data pairs to produce a meaningful plot.

2. Mark the horizontal (x) axis with the name of the first variable measured and its unit of measure. If there is a suspected cause-and-effect relationship between the variables, then the "cause" is normally plotted on this axis. This is the independent variable.

3. Mark the vertical (y) axis with the name of the second variable measured and its unit of measure. In cause-and-effect relationships, this is the suspected "effect," or dependent variable.

4. Divide and label the horizontal and vertical axes into even increments that easily allow all the data to be plotted. Both axes should be approximately of equal length so that a square plot results. Generally, the scale begins with 0, except where the range will not conveniently fit, in which case the scale begins with the lowest value.

5. Plot the data points. (Plots on log and semi-log paper often illustrate relationships that are not apparent on regular graph paper.)

6. Optionally, grid lines or an approximated regression line may be added to the graph if they improve clarity or aid in understanding the data.

Interpretation

1. If the resulting plot resembles a near horizontal or near vertical line, no relationship between the variables exists, and the visual appearance of correlation may be due to improper scaling of the axes. Similarly, if the resulting plot resembles a shotgun random pattern, again no relationship exists.

2. If the resulting plot has a definite slope and appears to be grouped in a non-random pattern, then a relationship (either positive or negative, linear or curved) may exist between the variables. For these plots, test the correlation of the Scatter Diagram to determine if further analysis is warranted. (Calculating the correlation coefficient and performing regression analysis are beyond the scope of this book, however.)

   • Count the data points on the chart if the number of pairs (n) is not known.

   • Draw median lines that separate the data points in half on both the x and y axes.

- Label the four quadrants. Start at the upper right and moving clockwise label them I, II, III, and IV.

- Count the data points in each quadrant.

- Add the data points in opposing quadrants (i.e., I + III and II + IV) and select the lower sum.

- Compare this sum with the values in Table 10.5 for the appropriate sample size (n). Only sample sizes from 30 to 79 are shown. If the number of points in the opposing quadrants is equal to or smaller than the test value, a correlation exists. (For n values not listed, use the test value for the next lower sample size. For example, the test value for 32 is 9, the same as n = 30.)

TABLE 10.5   Scatter Diagram Correlations

Sample Size (n)	Test Value: Lowest Two Quadrants	Sample Size (n)	Test Value: Lowest Two Quadrants
30	9	56	20
33	10	58	21
35	11	61	22
37	12	63	23
40	13	65	24
42	14	67	25
44	15	70	26
47	16	72	27
49	17	74	28
51	18	77	29
54	19	79	30

STATISTICAL PROCESS CONTROL: CONTROL CHARTS

Description

Control Charts are a specialized form of Run Charts with statistically determined upper and lower control limit lines added. The purpose of a Control Chart is to classify probable causes of process variation into either common or special causes. These control limits help to determine if the process remains stable or if something has changed. Before Control Charts can be applied, the process must be in a state of statistical control. This "in control" condition is discussed in the section titled "General Construction Steps," below.

There are seven main types of Control Charts; each has a particular usefulness, depending on the type of data collected:

- $\overline{X}$-R, x-R, and $\overline{X}$-s for variable (continuous interval) data

- x-R, np, p, c, and u for attribute (discrete interval) data

Statistical Process Control (SPC) provides a means to control processes, but it is not intended to be a substitute for inspection or testing and it is not a cure-all. SPC has limited applications, for example, when a process contains short and small runs (note that short runs may not necessarily be small runs). In these situations, there may be limited data, often not enough data to make a Histogram or a Control Chart. These situations require the use of modified SPC techniques designed for short and small runs (see Appendix B: References and Resources, SPC). Cumulative probability plots often provide insight into processes where there is limited data and where the real information is in the data that is not the process's normal output (see Appendix B: References and Resources, SPC).

One last caution before using Control Charts: not every element in a process needs to be monitored. Process stakeholders who understand the process generally know the vital few process characteristics that need to be measured. Key areas to explore include generic process elements that affect many critical steps, failure points, and elements sensitive to wide fluctuations in time and resources. A thorough review of Flowcharts often highlights these critical areas that justify the use of Control Charts.

Key Points

- Indicates whether or not a process is stable (i.e., in control)

- Provides a common language for comparing process performance

- Processes in control demonstrate variation that is natural (common cause)

- Variation outside of control limits (out of control) or statistically improbable trends are due to special causes

- Processes can be in control but incapable of meeting specifications

- Graphically illustrates when special causes that require intervention occur and when common causes occur, where no action is required to maintain the process

- Applied to interval data only (either continuous or discrete) and not to ordinal (i.e., sequential data) or nominal data (i.e., named categories)

- Different types of Control Charts are used for different types of interval data, either continuous or discrete; the x-R is used for both types

Typical Applications

- Determine whether or not the current process, either mechanical or manual, is capable of meeting expectations or specifications

- Determine if the process is in control (i.e., stable and predictable) and varies within expected limits

- Decide whether the process has changed due to special causes and will require adjustments to perform normally

- Provide a baseline of data for process improvement (i.e., a change in the process to alter common causes) in order to reduce variation and increase productivity by shifting the process mean

Example

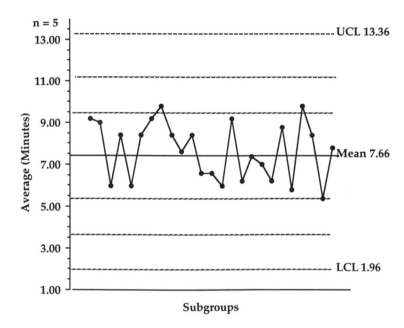

FIGURE 10.22   Control Chart: Video One Average Check-Out Time

Statistical Definitions

Attribute (Discrete) and Variable (Continuous) Data

Attribute data is obtained by counting non-measurable characteristics. This type of data consists of integral values or whole numbers. Examples of attribute data include any data that is *based on counting* discrete events (attributes) such as frequency of defects or number of defective items.

Variable data is based on *measurements* expressed as fractions or with decimals. Examples include length in feet, speed in miles per hour, weight in grams, volume in liters, and time in minutes. Variable data can, however, be expressed as attribute data; an example of converting variable data to attribute data would be to record length as either acceptable or not acceptable.

Common and Special Cause Variability

Common cause variation is the random variation that is inherent in the process itself. Variation due to common causes cannot usually be reduced unless the *process itself is changed.*

Because most process measurements are expected to fall within three standard deviations (see definition below) of the mean, the total spread of the process measurement is six standard deviations. Data that do not fall within the spread, or which fail to follow statistical probability rules, are said to be due to special causes.

Special cause variation is the variation in a process that can be assigned to a specific cause (sometimes referred to as assignable cause). Special cause variation is unpredictable, occasional, and unstable. When special cause variation is signaled by the chart, a search for the cause must begin. Once the special cause is identified, steps must be taken to prevent recurrence (if the change was unfavorable) or to ensure continuance (if the change was favorable). Fundamental process changes should not be made while special cause variations are occurring.

Treatment of common cause variability as a special cause is referred to as tampering. Tampering results in added process variability, higher costs, fear and poor morale, and wasted effort.

Mean

The mean, or average, of a data set is the total of all values divided by the number of values totaled. It is commonly designated as x-bar ($\overline{X}$). ($\overline{X}$ refers to the average of sample data, while μ is the notation used to represent the population average. Because Control Charts use sample data, the $\overline{X}$ notation is used.)

Normal Distribution

A normal distribution is a symmetrical, bell-shaped distribution in which the mean, median, and mode are all the same. Further, the total area under the curve is set equal to one, and the area under any portion of the curve can be calculated and converted to probabilities. A normal curve is illustrated (Figure 10.23) along with a Histogram generated from data with a normal distribution.

Measurements from a stable process usually will exhibit a central tendency that results in a normal curve whose mean approximates the population mean. For example, sample averages will reflect a normal distribution and approximate the population mean for any shape of population distribution, provided the sample size is sufficient (i.e., sample size exceeding 25).

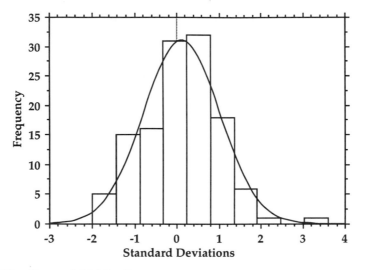

FIGURE 10.23   Normal Distribution

Range

Range is a measure of the variability that exists in a data set. It is equal to the lowest value in a data set subtracted from the highest value.

Standard Deviation

Standard deviation is a measure of the variability that exists in a data set and is commonly represented by the lowercase Greek letter sigma (σ). In a normal distribution, 68.26% of the data will be within one σ from the mean, 95.45% within two σ, and 99.73% within three σ. Mathematically, the standard deviation (see formula below) is equal to the square root of the average squared difference between individual data values and the data set average.

The use of σ to represent the standard deviation when referring to Control Charts has become common practice and, therefore, will be used here. This is not totally accurate, however; σ refers to the population standard deviation while s, or $\hat{\sigma}$, refers to the sample standard deviation. Control Charts use sample data and should use the sample standard deviation notation. The next section on Process Capability does use $\hat{\sigma}$ to indicate the sample standard deviation.

$$\sigma = \sqrt{\frac{(X_1 - \overline{X})^2 + ... + (X_n - \overline{X})^2}{(n - 1)}}$$

Control Chart Types

The seven main types of Control Charts are $\overline{X}$-s, $\overline{X}$-R, x-R, np, p, c, and u. The $\overline{X}$-s chart will not be covered here. (The s chart is the sample standard deviation chart and is beyond the scope of this book.) A brief description of the commonly used charts is provided here to help in understanding the differences between them; complete details for their construction are included later.

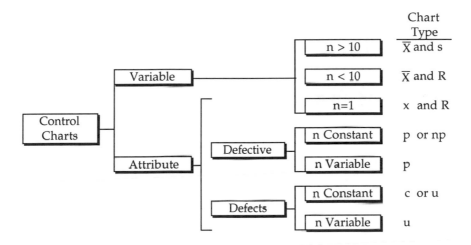

FIGURE 10.24   Control Chart Types

Variable Data Control Charts

Variable data require that two Control Charts be used in pairs: one to monitor the process average and the other to monitor process variation. Attribute charts, on the other hand, require only one chart.

$\overline{X}$-R

The $\overline{X}$-R Control Chart is used for variable data and is not used for attribute data. The $\overline{X}$ chart plots the mean of the data values, while the R chart plots the range of the data values. The sample size, n, is usually 10 or less (i.e., 10 data observations per sample).

x-R

This type of chart is often referred to as an individuals or moving range Control Chart. The x-R Control Chart is very similar to the $\overline{X}$-R Control Chart in that it is used for variable data and is actually a pair of charts. The x chart plots the individual data measurements (rather than the data means plotted in the $\overline{X}$ chart), while the R chart plots the range *between* data values (i.e., between the current and previous data values). Unlike the $\overline{X}$-R chart, however, the x-R chart can be and often is used with attribute data.

The primary usefulness of x-R charts is in situations where Sampling is either very costly, very time-consuming, or infrequently performed. It is common in batch-type processes where some variables continuously change and where a single measure provides current data, such as a temperature reading. The main disadvantage of the x-R chart is that it does not detect process changes as rapidly as the $\overline{X}$-R chart does.

Attribute Data Control Charts

The distinction between defective and number of defects must be understood before using attribute Control Charts. Defective items are rejected because they do not meet requirements, whereas items with defects or flaws are not automatically rejected because that specification may allow a certain level of defects.

np

The np Control Chart is used to monitor the *number of defectives in a sample of constant size.* It is used only with attribute data and is not applicable to variable data. Typical examples include monitoring the number of defective cars produced per day or incorrect paperwork redone per day.

p

The p Control Chart is used to monitor the fraction or *percentage defective in a sample of either varying or constant size.* Control limits vary if the sample size varies. It is used only with attribute data and is not applicable to variable data. Typical examples include monitoring the fraction (or percentage) of defective production per hour or the percentage incorrect paperwork per hour.

c

The c Control Chart is used to monitor the *count* of defects or flaws (non-conformities) for a *sample size that is constant.* It is used only with attribute data and is not applicable to variable data. Typical examples include monitoring the number of errors in a 10-page form or the number of defects in a part or assembly.

u

The u Control Chart is a more general form of the c chart. It is used to monitor the *count* of defects or flaws (non-conformities) in a *sample size that is either varying or constant.* It is used only with attribute data and is not applicable to variable data. Typical examples include monitoring the number of defects found in an hourly sample or the number of clerical errors in the forms reviewed daily.

The np Control Chart plots values that do not exceed the sample size since only a proportion of the total can be rejected. The c and u charts, however, plot the number of defects, which can exceed the sample size.

The difference between the np and c charts illustrates the difference between the number of defects and defectives. If 10 documents were inspected and 3 of the documents contained a total of 6 errors while the other 7 were perfect, an np chart would plot a 3 for the 3 defectives *if the specification allowed no errors.* A c chart would plot the total number of errors in all the documents sampled and plot a 6. The c chart would not indicate how many documents will be rejected.

General Construction Steps

1. Determine the type of data that is to be collected. For variable, or continuous, data, an $\overline{X}$-R or x-R chart is needed. For data that are attribute, or discrete (counting or pass/fail) in nature, a decision must be made as to whether an x-R, np, p, c, or u chart is best for the application.

2. Collect the data. See the specific construction steps below.

3. Mark the vertical (y) axis with the name of the attribute or variable measured and its unit of measure. Variable data charts will show two measures: the averages and the ranges.

4. Mark the horizontal (x) axis with how the process is divided up for measurement purposes and its unit of measure (e.g., subgroup number or measurement number).

5. Label the chart with meaningful title(s) which describe the process, steps in the process, unit of measure, dates, and other pertinent information.

6. Divide and label the horizontal and vertical axes into even increments which easily allow all the data to be plotted.

7. Calculate upper and lower control limits, and calculate the data mean and range as appropriate.

   A process is out of control if three or more values used to calculate the control limits exceed those limits. If two or less values exceed the limits, drop the variant samples and recalculate the control limits. If any remaining sample values fall outside of the limits, the process is out of control. All out-of-control conditions (that is, special causes) need to be corrected before controls can be established.

8. Plot the data points and the control limits.

9. Optionally, data point labels may be added to the graph if they improve clarity or aid in understanding the data.

Specific Construction Steps

$\overline{X}$-R Charts

The $\overline{X}$-R Control Chart is actually a pair of Control Charts. The $\overline{X}$ chart plots the mean of the data values and primarily shows changes that occur in the process average, while the R chart plots the range of the data values and primarily shows process variability.

The steps for creating an $\overline{X}$-R chart are:

1. Collect and record the data. Typically, more than 125 samples, or data points, are needed. They should be relatively recent and from a process that will be used in the future.

2. Divide the data into logical subgroups. A minimum of 25 subgroups is necessary to calculate control limits. Subgroups should consist of data drawn from the same lot or under similar conditions. For example, samples collected on the same date, time, or lot and under the same conditions can form a subgroup. They typically are in the range of two to five data points each to simplify calculations. These subgroups define the sample size, n.

3. For each subgroup, calculate its mean value, $\overline{X}$, and its range, R (i.e., the average of the data values in each subgroup and the difference between the highest and lowest data value in each subgroup). Calculate $\overline{X}$ to one decimal place beyond the sample measurement.

4. Calculate the mean of all the data points ($\overline{\overline{X}}$). This can be done by either averaging all the individual data points or by averaging the subgroup means, $\overline{X}$, calculated in step 3 above. Calculate $\overline{\overline{X}}$ to two decimal places beyond the sample measurement. This number becomes the centerline of the $\overline{X}$ chart.

5. Calculate the average range ($\overline{R}$). This is calculated by averaging all the ranges (R) calculated in step 3. Calculate $\overline{R}$ to one decimal place beyond the sample measurement. This becomes the centerline of the R chart.

6. Calculate the control limit lines for the $\overline{X}$ and R charts.

   Mathematicians have developed factors for Control Charts that approximate a $\pm 3\sigma$ variation for each possible number of data points in a subgroup of sample size n. To use these factors, it is necessary to know the mean of all data points ($\overline{\overline{X}}$) and the average range ($\overline{R}$) as previously calculated in steps 4 and 5.

   The factors listed in Table 10.6 allow calculation of the upper and lower control limits (UCL and LCL) for the $\overline{X}$ and R charts for sample sizes (n) up to ten. Note that for subgroups of six or less, the lower control limit for the R chart (LCL_R) is not applicable and no calculation is required. Although other reference sources list factors for subgroups larger than ten,

TABLE 10.6 Factors for $\overline{X}$ and R Charts

Data Points in Subgroup	Factors for $\overline{X}$ Chart	Factors for R Chart Upper Limit	Factors for R Chart Lower Limit
n	A2	D4	D3
2	1.880	3.267	N/A
3	1.023	2.575	N/A
4	0.729	2.282	N/A
5	0.577	2.115	N/A
6	0.483	2.004	N/A
7	0.419	1.924	0.076
8	0.373	1.864	0.136
9	0.337	1.816	0.184
10	0.308	1.777	0.223

an $\overline{X}$-s chart should be used in such cases instead of the $\overline{X}$-R. When the sample size exceeds ten, the range measurement is a biased estimate of variability and the unbiased sample standard deviation provides a more accurate measure of variability.

Calculate the upper and lower control limit lines (UCL and LCL) for the $\overline{X}$ chart as follows (calculate control limits to two decimal places beyond the sample measurement):

$$UCL_{\overline{X}} = \overline{\overline{X}} + A_2 \overline{R}$$
$$LCL_{\overline{X}} = \overline{\overline{X}} + A_2 \overline{R}$$

Calculate the upper and lower control limit lines (UCL$_R$ and LCL$_R$) for the $\overline{R}$ chart as follows (calculate control limits to one decimal place beyond the sample measurement):

$$UCL_R = D_4 \overline{R}$$
$$LCL_R = D_3 \overline{R}$$

Steps 7 through 11 require either graph paper or Control Chart forms (see Appendix B: Resources and References).

7. Mark the vertical (y) axis with the name of the average and the range variables measured and their units of measure.

8. Mark the horizontal (x) axis with how the process is divided up for measurement purposes and its unit of measure (e.g., subgroup number or measurement number).

9. Label the charts with meaningful title(s) and other pertinent information that describes the process and identifies the step in the process, the unit of measure, the period when it was taken, etc.

10. Divide and label the horizontal and vertical axes into even increments which allow all the data to be plotted. Draw in all control lines and label them with numerical values:

    • Solid lines for the process average, $\overline{\overline{X}}$, and the average range, $\overline{R}$

    • Broken lines for the control limits

      Indicate the subgroup, or sample size, n, on the left side and above the $UCL_{\overline{X}}$

11. Plot the values for each subgroup using a dot (·) for $\overline{X}$ values and an x for R values. Circle all points outside of the control limits.

Example

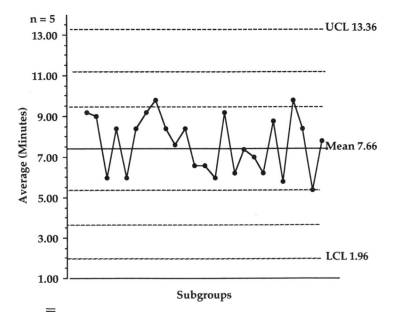

FIGURE 10.25 $\overline{X}$ **Control Chart: Video One Average Check-Out Time**

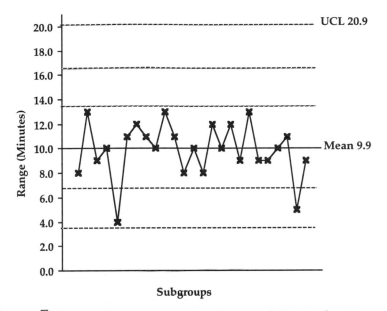

FIGURE 10.26 R̄ Control Chart: Video One Range of Check-Out Times

X̄-R Variations

The X̄ chart is sometimes used without its R chart complement. This is done in situations where the variation in the data is small or of less concern.

x-R

This type of chart is often referred to as an individuals or moving range Control Chart. The x-R Control Chart is very similar to the X̄-R Control Chart in that it can be used for variable data and is actually a pair of charts. The x chart plots the individual data measurements (rather than the data means plotted in the X̄ chart), while the R chart plots the range *between* data values (i.e., between the current and previous data values) and primarily shows process variability. The main disadvantage of the x-R chart for variable data is that it does not detect process changes as rapidly as the X̄-R chart does.

Unlike the X̄-R, however, the x-R can be and often is used with attribute data. Virtually all count (attribute) data can be plotted using the x-R chart; only when the average count of defects or defectives is two or less per sample is the x-R statistically invalid. The np, p, c, and u charts are aimed at specific types of count data. They offer greater reliability and accuracy than the more generic x-R chart for most count data.

The primary usefulness of x-R charts is in situations where Sampling is very costly, very time-consuming, or infrequently performed. It is common in batch-type processes where some variables continuously change and where a single measure provides current data, such as a temperature reading.

The steps for creating an x-R chart are:

1. Collect and record the data. Typically, more than 25 values are needed. They should be relatively recent and from a process that will be used in the future.

2. Calculate the mean of all the data points ($\overline{X}$). This number becomes the centerline of the x chart.

3. Calculate the range *between* each data point (i.e., the absolute value of the difference between the current data point and the previous data point). Note that there is no range value for the first data point, and therefore the chart will always have one fewer range value than there are individual measurements. Also note that, as in the $\overline{X}$-R chart, range is by definition a positive number.

4. Calculate the average range ($\overline{R}$). This is calculated by averaging all the ranges ($\overline{R}$) calculated in step 3. This becomes the centerline of the R chart.

5. Calculate the control limit lines for the x and the R charts.

   Mathematicians have developed a factor (2.66) for the x Control Chart that approximates a 3σ variation. To use this factor, it is necessary to know the mean of all data points ($\overline{X}$) and the average range ($\overline{R}$) as previously calculated in steps 2 and 4.

   Calculate the upper and lower control limit lines (UCL_X and LCL_X) for the x chart as follows:

   $$UCL_X = \overline{X} + 2.66\,\overline{R}$$

   $$LCL_X = \overline{X} + 2.66\,\overline{R}$$

   If a $\pm 2\sigma$ variation is desired (instead of the $\pm 3\sigma$ variation provided by the above equations), 2.66 should be replaced by 1.77.

   Calculate the upper control limit line (UCL_R) for the R chart next. Use Table 10.6 and the average range ($\overline{R}$, as previously calculated in step 4) to calculate the upper control limit for the R chart using the formula below. Note that the subgroup size (n) for the R chart of the x-R is always two and the lower control limit line (LCL_R) is not applicable. No calculation is required.

   $$UCL_R = D_4\,\overline{R}$$

   This simplifies to the following formula:

   $$LCL_R = 3.267\,\overline{R}$$

   Steps 6 through 10 require either graph paper or Control Chart forms (see Appendix B: Resources and References).

6. Mark the vertical (y) axis with the name of the variable measured and its unit of measure.

7. Mark the horizontal (x) axis with how the process is divided up for measurement purposes and its unit of measure.

8. Label the charts with meaningful title(s) and other pertinent information that describes the process and identifies the step in the process, the unit of measure, the period when it was taken, etc.

9. Divide and label the horizontal and vertical axes into even increments which allow all the data to be plotted. Draw in all control lines and label them with numerical values:

   • Solid lines for the process average, $\overline{X}$, and the average range, $\overline{R}$

   • Broken lines for the control limits

10. Plot the values for each subgroup using a dot (•) for x values and an x for R values. Circle all points outside of the control limits.

np and p

The np Control Chart is used to chart the *number of defectives in a sample of constant size*. The p Control Chart is used to chart the fraction or *percentage defective in a sample of either varying or constant size*. Both charts are used only with attribute data and are not applicable to variable data. Unlike the $\overline{X}$-R or x-R Control Charts, the np and p Control Charts are not used in combination with each other or another chart because both the characteristics of the mean and variability are shown on the same chart.

Typical np Control Chart examples include charting the number of defective cars produced per day or incorrect paperwork redone per day. Typical p Control Chart examples include charting percentage incorrect forms per hour or fraction defective production per hour.

The steps for making a p chart are:

1. Collect and record the data. Divide the data into subgroups by date or lot, for example. Generally, there should be 25 subgroups with a subgroup size (n) that is typically greater than 50, where the average number of defective (np) found in each data pair should be 3 to 4 or greater. The data should be relatively recent and from a process that will be used in the future.

2. For each subgroup, calculate the fraction defective (p). This is simply the number of defectives (np) divided by the number inspected in the sub-group (n). To convert to the percentage defective, multiply each result by 100.

3. Calculate the centerline for the chart: the average fraction defective ($\overline{p}$). This is the grand total of all defectives divided by the grand total of all inspected.

$$\overline{p} = \text{total defective} \div \text{total inspected} = \Sigma np \div \Sigma n$$

4. Calculate the control limits as follows:

$$UCL_p = \overline{p} + 3\sqrt{\frac{\overline{p}(1 - \overline{p})}{n}}$$

$$LCL_p = \overline{p} - 3\sqrt{\frac{\overline{p}(1 - \overline{p})}{n}}$$

A negative result for a calculated lower control limit for the p chart indicates that no lower control limit can be plotted.

Note that as the number inspected (n) changes, the control limits change. **This means that a new upper and lower control limit must be calculated for _every_ subgroup.** As a simplification, where all the numbers inspected (n) are within 20% of the average sample size, then the average sample size ($\overline{n}$) may be substituted for all subgroups (i.e., substitute $\sqrt{\overline{n}}$ for $\sqrt{n}$ in the control limit formulas). This simplification results in straight line control limits. Calculate individual limits for samples exceeding ±20%.

5. Label the Control Chart with meaningful titles and other pertinent information. Label the x and y axes and draw a solid line for the mean fraction defective ($\overline{p}$). Use broken lines to identify the control limits. Plot the fraction defective (p) with appropriate control limits based on the varying sample size (n).

The steps for making an np chart are:

1. Collect and record the data. Divide the data into subgroups. Generally, there should be 25 subgroups with a constant subgroup size (n) that is typically greater than 50, where the average number of defectives (np) found should be 3 to 4 or greater. The data should be relatively recent and from a process that will be used in the future. Note that the np chart really does not require a data pair because the number inspected (n) remains constant.

2. Calculate the centerline for the chart: the average number defective per subgroup ($n\overline{p}$). This is the grand total of all defectives divided by the number of subgroups inspected.

$n\overline{p}$ = total defective ÷ number of subgroups = Σnp ÷ number of subgroups

3. Calculate the control limits as follows, where $\overline{p}$ is the average fraction defective (i.e., $n\overline{p}$ ÷ n):

$$UCL_{np} = n\overline{p} + 3\sqrt{n\overline{p}(1 - \overline{p})}$$

$$LCL_{np} = n\overline{p} - 3\sqrt{n\overline{p}(1 - \overline{p})}$$

A negative result for a calculated lower control limit for the p chart indicates that no lower control limit can be plotted.

4. Label the Control Chart with meaningful titles and other pertinent information. Label the x and y axes and draw a solid line for the mean fraction defective ($n\bar{p}$). Use broken lines to identify the control limits. Plot the number of defectives (np).

c and u

The c Control Chart is used to chart the *count* of defects or flaws (non-conformities) for a *sample size that is constant*. It is used only with attribute data and is not applicable to variable data. Typical examples include charting the number of errors in a 10-page form or the number of defects in a part or assembly.

The np Control Chart plots values that do not exceed the sample size because only a proportion to the total can be rejected. The c and u charts, however, plot the number of defects, which can exceed the sample size.

The difference between the np and c charts illustrates the difference between the number of defects and defectives. If 10 documents were inspected and 3 of the documents contained a total of 6 errors while the other 7 were perfect, an np chart would plot a 3 for the 3 defectives *if the specification allowed no errors*. A c chart would plot the total number of errors in all the documents sampled and plot a 6. The c chart would not indicate how many documents will be rejected.

The u Control Chart is a more general form of the c chart. It is used to chart the *count* of defects or flaws (non-conformities) in a *sample size that is either varying or constant*. It is used only with attribute data and is not applicable to variable data. Typical examples include charting the number of defects found in an hourly sample or the number of clerical errors in the forms reviewed daily.

The steps for making a u control chart are:

1. Collect and record the data. Typically, at least 25 pairs of data will be needed, which are grouped by lot, products, samples, etc. Each pair includes the number inspected, that is, the subgroup size (n) and the number of defects found (c). The data should be relatively recent and from a process that will be used in the future. The number inspected (n) should typically be greater than 50 and the average number of defects found in each data pair should be greater than 2 or 3.

2. For each subgroup, calculate the ratio (u) of defects to number inspected (n). This is simply the total number of defects (c) divided by the number of units inspected (n). Because the subgroup size varies, a value for the defects per sample (u) is calculated for each subgroup.

3. Calculate the centerline for the chart: the average number of defects ($\bar{u}$). This is the grand total of all defects divided by the grand total units in all subgroups.

$$\bar{u} = \text{total defects} \div \text{total inspected} = \Sigma c \div \Sigma n$$

4. Calculate the control limits as follows:

$$UCL_u = \bar{u} + 3\sqrt{\frac{\bar{u}}{n}}$$

$$LCL_u = \bar{u} - 3\sqrt{\frac{\bar{u}}{n}}$$

A negative result for a calculated lower control limit for the u chart indicates that no lower control limit can be plotted.

Note that as the number inspected (n) changes, the control limits change. **This means that a new upper and lower control limit must be calculated for *every* subgroup.** As a simplification, where all the numbers inspected (n) are within 20% of the average sample size, then the average sample size ($\bar{n}$) may be substituted for all subgroups (i.e., substitute $\sqrt{\bar{n}}$ for $\sqrt{n}$ in the control limit formulas). This simplification results in straight line control limits. Calculate individual limits for samples exceeding ±20%.

5. Label the Control Chart with meaningful titles and other pertinent information. Label the x and y axes and draw a solid line for the average number defective ($\bar{u}$). Use broken lines to identify the control limits. Plot the number of defects per sample (u) with appropriate control limits based on the varying sample size (n).

The steps for making a c Control Chart are:

1. Collect and record the data. Divide and group the data by lot, products, samples, etc. Generally, there should be 25 subgroups with a constant subgroup size (n) that is greater than 50, where the average number of defects (c) found in each subgroup should be greater than 2 or 3. The data should be relatively recent and from a process that will be used in the future. Note that the c chart really does not require a data pair because the number inspected (n) remains constant.

2. Calculate the centerline for the chart: the average number of defects ($\bar{c}$). This is the grand total of all defects divided by the number of subgroups inspected.

$$\bar{c} = \text{total defects} \div \text{number of subgroups} = \Sigma c \div \text{number of subgroups}$$

3. Calculate the control limits as follows:

$$UCL_c = \bar{c} + 3\sqrt{\bar{c}}$$

$$LCL_c = \bar{c} - 3\sqrt{\bar{c}}$$

A negative result for the lower control limit of the c chart indicates that no lower control limit can be plotted.

4. Label the Control Chart with meaningful titles and other pertinent information. Label the x and y axes and draw a solid line for the average

number defective ($\overline{c}$). Use broken lines to identify the control limits. Plot the number of defects (c).

Interpretation

Most of the variation in a Control Chart is due to random variation in the process itself (common causes) and requires no adjustments, or tampering, unless there is a desire to improve the process (i.e., to improve the mean and reduce the variation, which should be a continuous goal).

Only some of the variation shown on a Control Chart is due to unusual circumstances (special causes) and needs special attention (i.e., analysis and correction). Variations that are statistically improbable are said to be due to special causes. Clearly, variations outside of the control limits are due to special causes. In addition, patterns of variations *within the control limits* can also signal the existence of a special cause that requires correction. These variations generally can be associated with one of several categories, such a people, machines, materials, methods, measurements, and environment. The following rules are based on the statistical probabilities of various patterns that occur in the data:

1. It is statistically improbable that a point outside of the control limits is due to random variation (common causes). There is either 95.45 or 99.73% probability (i.e., limits based on 2 or 3 standard deviations, respectively) that the variation is due to a special cause. A data point that exceeds the upper control limit is illustrated in Figure 10.27.

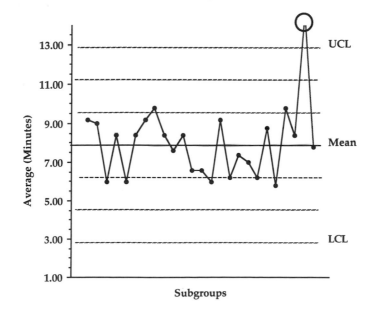

FIGURE 10.27 Special Cause

2. A process is changing or not stable when at least two out of three successive points on the same side of the centerline are at least two standard deviations (2σ) away from the process centerline.

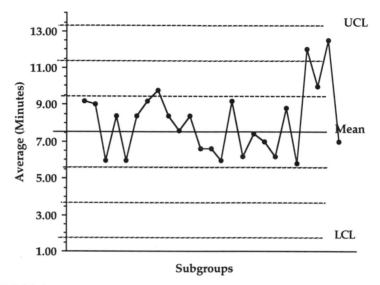

FIGURE 10.28   Two of Three Points Greater Than Two σ from Centerline on Same Side

3. A process is changing or not stable when at least four out of five successive points are more than one standard deviation (1σ) away from and on the same side of the centerline.

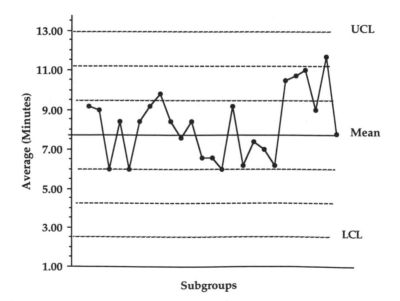

FIGURE 10.29   Four of Five Points Greater Than One σ from Centerline on Same Side

4. One should expect to see an equal number of points above and below the process centerline. Whenever nine or more points in a row remain ("run") on one side of the centerline, this indicates that a process change has occurred.

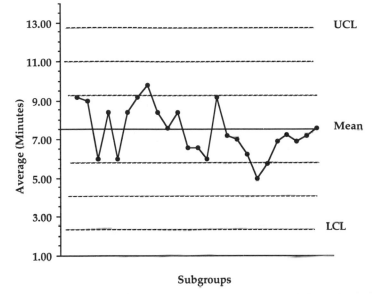

FIGURE 10.30   Run of Nine Points All Below Centerline

5. A trend of six or more points *continuously* increasing (or decreasing) also indicates that the process is changing. A Control Chart with six points increasing is illustrated in Figure 10.31; note that this is an example of six, not seven, points increasing. The first point that starts the trend is not counted.

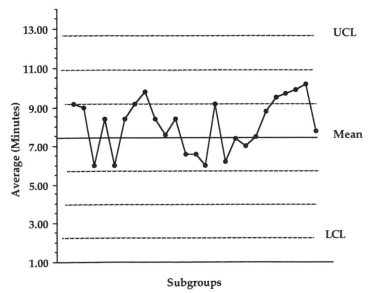

FIGURE 10.31   Six Points Continuously Increasing

6. Fourteen consecutive points alternating up and down is statistically improbable. A chart with fourteen points alternating up and down is illustrated in Figure 10.32; note that this is an example of *fourteen,* not fifteen, points alternating. The first point that starts the trend is not counted.

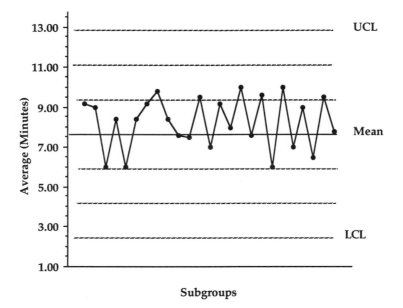

FIGURE 10.32   Fourteen Consecutive Points Alternating Up and Down

7. Fifteen successive points within one standard deviation (1σ) of the data centerline is statistically improbable and may indicate either data tampering or a drastic improvement in the process.

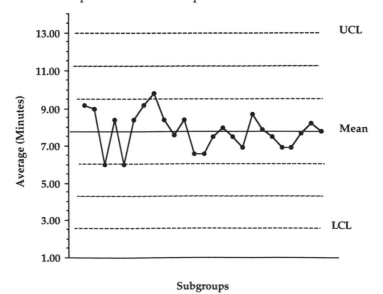

FIGURE 10.33   Fifteen Consecutive Points Within One σ of Centerline

8. A point that changes by four or more standard deviations (4σ), a jump, is statistically improbable.

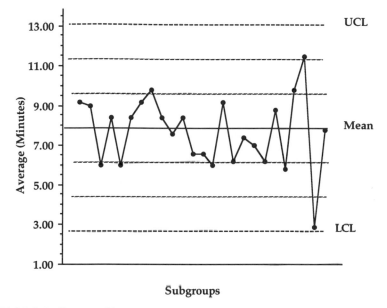

FIGURE 10.34   Four σ Change

STATISTICAL PROCESS CONTROL: PROCESS CAPABILITY

Description

Process Capability is a technique that is used to determine whether or not a process, given its natural variation, is capable of meeting customer specifications. Processes can be in control and produce outputs that do not meet customer needs and expectations. Process Capability uses an objective measure, a capability index, that compares process specification limits with the process control limits.

Specifications generally define a nominal value, or target measure, plus or minus an acceptable variation. These limits define the upper specification limit (USL) and lower specification limit (LSL). This becomes a two-sided specification. Some specifications may have a target value and only one specification limit, however.

The simple process potential index, C_p, compares the difference between the upper (USL) and lower specification limits (LSL) with the process variation. If the dispersion of the specifications (i.e., USL minus LSL) exceeds 6 times the estimated process standard deviation ($6\hat{\sigma}$), then the process can meet specifications, *given that the process average ($\overline{\overline{X}}$) is centered on the target value, or mid point, of the specifications.* The simple process potential index cannot be used if there is a one-sided specification.

The process capability index, C_{pk}, is preferred over the simple process potential index, C_p, because it compares both the process dispersion (i.e., UCL and LCL) with the specification limit (i.e., USL and LSL) and the process average ($\overline{\overline{X}}$) with the target value. C_{pk} is made up of two capability indices, a lower and upper capability (C_{pl} and C_{pu}). The smaller value of C_{pl} or C_{pu} determines the C_{pk}.

Key Points

- Processes that are in control do not necessarily produce product that meets customer specifications

- Simple process potential index, C_p, compares the dispersion of two-sided specifications with the variation of the process

- The process capability index, C_{pk}, or process capability, compares both dispersion (either two-sided or one-sided) and target values with the process variation and average

- Process capability index, C_{pk}, is composed of an upper and lower capability, C_{pl} and C_{pu}

- If $C_{pl} = C_{pu}$, then the process is exactly centered on the target value

Typical Applications

- Determine whether the process is capable of meeting customer specifications

- Negotiate realistic and meaningful specifications

- Negotiate supplier certification targets (e.g., $C_{pk} = 1.33$ for certification and $C_{pk} = 2.0$ for a long-term target)

Example

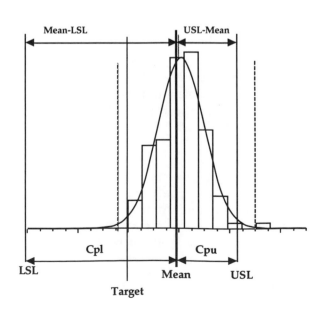

FIGURE 10.35
Process Capability: C_{pk} Chart

Steps

1. If the process average is aligned with the target value of the specification, the simple process potential, C_p, can be used for two-sided specifications. Calculate C_p using the estimated process standard deviation, $\hat{\sigma}$, in the following equation:

$$C_p = \frac{USL - LCL}{6\hat{\sigma}}$$

   where $\hat{\sigma} = \dfrac{\overline{R}}{d_2}$

   $\overline{R}$ = average of subgroup ranges

   n = subgroup sample size

   d_2 is determined from Table 10.7

TABLE 10.7 Process Capability Factors

Sample Size (n)	d_2
2	1.128
3	1.693
4	2.059
5	2.326
6	2.534
7	2.704
8	2.847
9	2.970
10	3.078

2. If the process average is not aligned with the target value of the specification, calculate both the C_{pu} and C_{pl} indices as follows:

$$C_{pu} = \frac{USL - \overline{\overline{X}}}{3\hat{\sigma}}$$

$$C_{pl} = \frac{\overline{\overline{X}} - LSL}{3\hat{\sigma}}$$

   where $\overline{\overline{X}}$ = the process average.

   The smaller of the two indices, C_{pl} or C_{pu}, defines the process capability, C_{pk}.

Interpretation

1. *Given that the process average is aligned with the target value of the specification,* the simple process potential, C_p, can be used to determine Process Capability for two-sided specifications. When $C_p > 1$, the process variation is less than specification. If the specification dispersion is equal to or less than the $6\hat{\sigma}$, then $C_p = 1$ and $C_p < 1$, respectively, and specification rejects will equal or exceed process rejects.

2. In all other cases, use C_{pk}, which is the smaller of C_{pl} or C_{pu}. When $C_{pk} > 1$, the process variation is less than specification. If the specification dispersion is equal to or less than the $3\hat{\sigma}$ for the limiting side, then $C_{pk} = 1$ and $C_{pk} < 1$, respectively, and specification rejects will equal or exceed process rejects.

Six Sigma

This program, popularized by the Motorola Corporation, has a goal of reducing process variation such that the spread between process specifications is $12\hat{\sigma}$ and the process mean is centered on the specification target. This means that the specification limits are six sigma from the process mean. Even if the process mean shifts up to 1.5 sigma from the target, the probability that the process produces outside of specification is only 3.4 non-conforming events per million. Six Sigma translates into a $C_{pk} = 2$.

CHAPTER ELEVEN

PLANNING TOOLS

Five tools for planning quality improvement are presented in this chapter: Activity Network Diagram, Gantt Chart, Process Decision Program Chart, Storyboards, and Tree Diagram. These tools are useful in the planning phase of the Quality Improvement Model (Steps 1 through 4) and in tracking that implementation (Steps 5 through 7). Of the five tools, Gantt Charts, Storyboards, and Tree Diagrams are the most useful in helping teams create their project road map in Step 1. The scope of the project generally dictates which tool is needed. If a complex tool, such as the Activity Network Diagram, is required, the team should reassess the project scope to ensure that it is not too broad.

Except for the Gantt Chart and Storyboards, these tools are included in the seven "new" management tools. The Activity Network Diagram is used to develop a sequential plan for implementing improvements. It provides a comparative reference for gauging progress and details the interrelationships between plan tasks. This tool is usually used on complex projects with serious consequences that are the result of delays. Gantt Charts provide a less rigorous planning tool with wider application to typical quality improvement projects.

Process Decision Program Charts are used to create a detailed implementation plan that includes all conceivable unfavorable events and possible problems that may occur. Each of these "what ifs" has a planned countermeasure to ensure the success of the plan. This tool is often more complex than is required for creating the project road map in Step 1, but various elements of the tool can be applied in that step. Planning for potential project risks and appropriate countermeasures should be included in the project plan.

Storyboards provide an organized, visual summary that is useful in planning quality improvement projects, as well as documenting and communicating results. They highlight the logical steps planned by the team, and once the project is underway, they provide a visual display of key findings and recommendations.

179

Tree Diagrams are only referenced here because the description and details of this tool are included in Chapter Seven. In addition to planning complex tasks, this tool also has applications in defining processes and in some cause-and-effect analyses.

ACTIVITY NETWORK DIAGRAMS

Description

The Activity Network Diagram is a management and planning tool that is used to develop a time–sequential plan for implementing projects or improvements. These diagrams provide a comparative reference for gauging progress and detail the interrelationships between tasks. They are similar to, and derived from, the Program Evaluation Research Technique (PERT) and the Critical Path Method (CPM) techniques, where the focus is to identify sequence of activities, risks of delays, and critical paths that can cause delays. The detail required generally limits their use to complex projects or projects for which the financial and strategic consequences of delays are significant.

Activity Network Diagrams are derived from the early 1930s Matrix Diagram (or Harmony Graph) of Polish scientist Karol Adamiecki. The next evolution included PERT charts and CPM, which were developed almost simultaneously in the late 1950s in defense and commercial applications by the 1958 Polaris team (Lockheed, the Navy, and Booz Allen and Hamilton) and the 1956 duPont/ Remington Rand Univac team, respectively, for example.

Arrow Diagrams (one of the seven "new" management tools) are one of three methods of drawing Activity Networks Diagrams. In this form, an arrow represents the activity and nodes (circles) represent the start and finish of the activity. The more common format uses "activities on the node" (where activities are described within rectangular nodes), or Node Diagrams, where arrows indicating the sequence of events. This format will be used here. The final format, Precedence Diagram, allows for partial completion and the beginning of succeeding steps.

Key Points

- Aids in determining the time required for a plan and each of its tasks, particularly when there is little margin for failure and the plan is both complex and vital to the organization

- Identifies tasks that can be completed simultaneously or that may be unnecessary; highlights critical paths that determine earliest possible completion time

- Supports maintaining project timeliness by establishing the time frame in which each task must start

- Time relationships can also be shown using Gantt Charts, but sequential dependencies are less obvious with Gantt Charts

- Responsibility for each activity can be identified

Typical Applications

- Identify the logical sequence and timing of activities required to implement a project within time and budget requirements

- Determine resource allocations needed to remain on schedule

- Limit use of key resources (e.g., overtime, expediting, etc.) to activities on the critical path only

- Estimate completion dates for detailed tasks within a complex project

Example

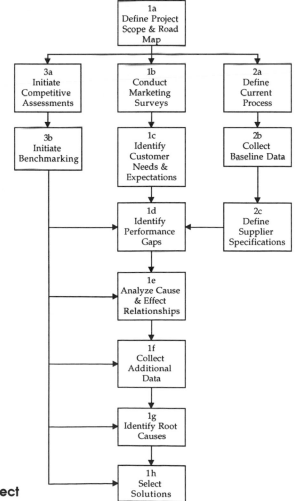

FIGURE 11.1
Activity Network Diagram:
Video One Check-Out Time Project

Steps

1. Generate a list of activities necessary to complete the plan. Start with higher level activities and add additional detail as needed. Brainstorming, existing Flowcharts, or other documentation of activities may be helpful in generating the list of activities.

2. Write each activity on the designated area of a "job card" (3 × 5 index card or 3 × 5 Post-It™ note) and a short description of the activity (see Figure 11.2). Use a verb-noun format. Departmental or individual responsibility for activities can be included above the activity description. (Note that Figure 11.1 shows summary level detail only.)

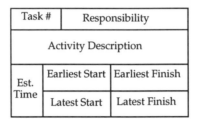

FIGURE 11.2   Sample Job Card

3. Sequentially arrange the cards into groups, or strings, of essential tasks. Work from left (early tasks) to right (subsequent tasks). Strings of activities that can be performed in parallel with other essential tasks should be separated into strings of simultaneous, or parallel, tasks. (Note that Figure 11.1 is shown in a vertical format rather than the horizontal one described here.)

4. Arrange the strings of activities by beginning at the top left and working toward the bottom right. Remove duplicate cards or add additional cards for more detail as the tasks are arranged. Parallel strings of simultaneous tasks are generally shown above the longest path. The first task of the parallel string is placed above (or below) the activity in the longest path when the parallel task can begin.

5. Once all activities are arranged in sequential order, assign a task number in the upper left corner of each job card. Work from the upper left to the bottom right.

6. Place the index cards or Post-It™ notes on butcher paper and draw arrows connecting strings of activities, including arrows to parallel tasks above (or below) the longest path.

7. Estimate how long it will take to complete each activity. Estimates can be based on historical performance or new estimates. Record this estimated time on the bottom left of each job card (Figure 11.2, "Est. Time").

8. Determine the critical path by finding which path has the longest completion time from start to end. The sum of times defines the project duration. The critical path is highlighted with bold lines.

9. Beginning with the first job card, or node, determine the earliest possible time each task could be started, given the time required for its predecessors. Record the earliest start time (ES) in the appropriate space on the job card (Figure 11.2). The estimated duration of each task added to the earliest start time determines the earliest finish time (EF), which in turn becomes the earliest start time for the next task in the sequence. Record these times in the appropriate spaces.

10. Beginning at the end node, or last job card, determine the latest possible time each task could be started and still finish the project on schedule. Record the latest start time (LS). The latest start time becomes the latest finish time (LF) of the preceding task.

11. The difference between the earliest start and latest start for each task is slack for that activity. The difference between the latest start of the critical path and the latest start of a parallel path is slack time for that path. By definition, the critical path has no slack time.

12. Review the Activity Network Diagram frequently as the plan is implemented. Adjust resources to ensure that all tasks are started as close to the earliest start time as possible and not later than the latest start time.

GANTT CHART

Description

The Gantt Chart is a planning tool that is used to document the schedule, activities, and responsibilities necessary to complete a project or implement a proposed solution. In addition to project planning applications, Gantt Charts are use in various applications where resources are allocated or scheduled. Although there are many forms and variations, all Gantt Charts identify what is to be accomplished, when it is to be completed, and who has responsibility for performing tasks.

Gantt Charts provide a visual picture of the sequence of activities, when each activity is proposed to begin and end, and actual progress relative to expected progress for a specific point in time. Periodic updates of progress portray current status against plan.

Key Points

- Demonstrates both the sequence and status of each task required to complete a plan

- Identifies tasks that can be completed simultaneously

- Has limited capability to show sequential dependencies and critical paths, key attributes of the Activity Network Diagram

- Supports maintaining project timeliness by establishing the time frame in which each task must start
- Responsibilities and assumptions can be identified for each activity

Typical Applications

- Identify the sequence and timing of activities required to implement a project within time and budget requirements
- Determine resource allocations needed to remain on schedule
- Project completion dates for tasks within a project

Example

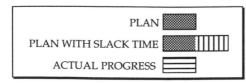

FIGURE 11.3 Gantt Chart: Video One Check-Out Time Project

Steps

1. Generate a list of activities necessary to complete the project. Break the plan into achievable tasks. Start with higher level activities and add

additional detail as needed. Brainstorming or other documentation of activities may be helpful in generating the list of activities.

2. Activities should be described in a verb-noun format. Assign specific departmental or individual responsibility for each task. Identify appropriate assumptions and required completion dates.

3. Group related activities and list them in the sequence in which they need to be performed. Tasks that are dependent on the completion of other tasks should be grouped together or identified in the assumptions.

4. Once all activities are listed, assign a task number to each task.

5. Estimate how long it will take to complete each activity. Estimates can be based on historical performance or new estimates. Use consistent nomenclature for the beginning, ending, and duration of the proposed activity, including slack time. Give careful attention to the sequence of dependent tasks to ensure that activities begin when preceding activities are completed.

6. Next, indicate the level of completion using consistent nomenclature. Because simple Gantt Charts do not show specific resource requirements and levels of completion, progress is estimated as a percent of the proposed task.

7. Draw a vertical date line to indicate the current date.

8. Review the Gantt Chart frequently as the plan is implemented. Adjust resources to ensure that all tasks are completed on schedule. Tasks that are on schedule will have an actual progress bar that coincides with the current time marker. All other tasks will be either ahead of or behind schedule.

PROCESS DECISION PROGRAM CHART

Description

The Process Decision Program Chart (PDPC) is a management and planning tool that is used to create a detailed implementation plan that includes all conceivable unfavorable events and possible problems that may occur. Identification of all feasible problems (i.e., those with a reasonable chance of occurring) before they occur enables development of contingency actions, or countermeasures. This tool is often used where complex tasks are being implemented for the first time and the cost of failure is high or the scheduled completion date is critical.

A PDPC recognizes that all activities toward a goal or objective must deal with an uncertain environment. A PDPC provides a low-cost method of testing process capability because it is a dry run, or pilot, of expected tasks. Its benefit is often validated in the study phase of the Quality Improvement Model (Step 6).

Key Points

- Often referred to as a contingency diagram

- Identifies what could go wrong with a plan or project (that is, a series of potential "what if" scenarios)

- Helps prevent situations or events from becoming problems by addressing potential countermeasures

- Most useful for plans that are new, unique, complex, and of great importance

Typical Applications

- Used to study plans in the conceptual stage under conditions of uncertainty to determine the need to revise plans prior to implementation

- Used to anticipate problems that have a reasonable expectation of occurring in plans that are complex, new, or have a high cost of failure or critical deadlines

- Often applied in selecting alternatives that have the least serious "what if" consequences of failure

Example

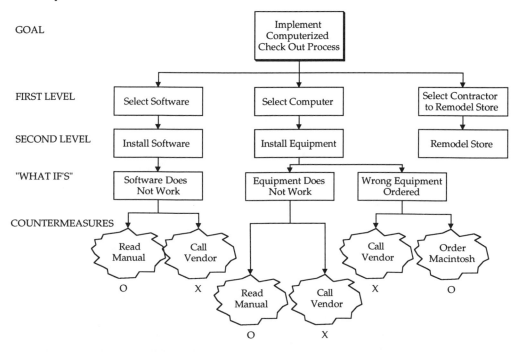

FIGURE 11.4   Process Decision Program Chart: Video One Check-Out Time Project

Steps

1. Generate a Tree Diagram (often with one or two layers of detail) for the project or plan. It is best to start at a low level of detail (i.e., where there are few branches to the tree) in order to avoid analysis of problems with insignificant consequences. Although there are two basic forms of the PDPC (graphic and outline format), the graphic format will be discussed here. The highest level in the Tree Diagram is the goal, which is then followed by the next level of detail in the implementation sequence (if the tree is horizontal, start at the left and work toward the right; if vertical, start from the top and work down).

2. Select an item from the lowest level of detail on the Tree Diagram. For each item, ask, "What potential problems could arise while performing this item?" or "What could go wrong here?" and brainstorm a list of answers. These are the "what ifs" for which countermeasures need to be developed. Review these responses and eliminate any that do not seem to be plausible. The objective is to identify potential problems that have a reasonable expectation of occurring.

3. List these "what ifs" as the next layer in the Tree Diagram. Brainstorm and list potential "what ifs" for all items.

4. Brainstorm possible countermeasures (i.e., actions to counter the problems anticipated) for each "what if." List all countermeasures as the next layer of detail in the Tree Diagram. Countermeasures should be enclosed in a cloud-shaped icon to distinguish them from other graphics. Connect countermeasures with appropriate "what ifs."

5. Evaluate the feasibility of each countermeasure. Those that are impossible, impractical, and/or difficult should be marked with an "X." Those that are possible, practical, and selected should be marked with an "O."

    (In the outline format, the outline of tasks is created using a numerical indexing system such that all first-level tasks are designated at the integer level, say 1.0, and second-level tasks at the first decimal level, say 1.1. The nomenclature (or nesting) for the "what if" situation at this second layer would be 1.1.1, and the countermeasures would be designated as 1.1.1.1 and 1.1.1.2, for example. These countermeasures would be preceded by and "X" and an "O," as in step 7 above. When the list is sorted using this index, the order is first-level task, second-level task, "what if," and then countermeasures.)

STORYBOARDS

Description

The Storyboard concept has been used in the entertainment industry (starting with cartoon animation in the 1930s) to portray the flow of a story in the planning stages. Storyboards provide an organized, visual summary that is useful in planning quality improvement projects as well as documenting and communicating results. They highlight the logical steps planned by the team, and once the project is underway, they visually display key findings and recommendations. When used to communicate project performance, they show the sequence of events and summary data, including process Flowcharts and other graphic tools that document the team's findings.

Storyboards and Meeting Minutes (Chapter Twelve: Meeting Management Tools) are the two primary tools used to document team efforts. Together they provide a road map that others can follow in related studies (i.e., related in content or team process) or to replicate the project in the future.

In addition, Storyboards can be used during project planning to generate and organize tasks. In this application, development of a detailed plan follows steps similar to those used to create Affinity Diagrams (Chapter Six) and then organize them into Tree Diagrams (Chapter Seven).

Key Points

- Provide an excellent means of planning, documenting, and communicating the improvement process followed by a team

- Help generate enthusiasm for others to apply continuous improvement

- Only summary data are portrayed, often in miniature form

- When used to generate and organize ideas, Storyboards are similar to Affinity and Tree Diagrams

Storyboard Rules

1. Do not reduce the size of text and graphics so that they are unreadable.

2. Be creative.

3. Storyboards used to document or communicate results of quality improvement should provide answers to the following questions:

    - What logical approach was followed?

    - What was the problem or opportunity statement and the scope of the project?

- Who was involved?

- Was the process being analyzed described in words or illustrated graphically?

- How were the cause-and-effect relationships analyzed?

- What other analytical tools were used?

- What were summary findings and recommended solutions?

- If the project is in implementation, how was the implementation planned, how were results monitored, and how was the solution standardized throughout the organization?

Typical Applications

- Plan the steps of a quality improvement activity

- Communicate the results of quality improvement

- Document the results of quality improvement for use as a guide to replicate the study in the future

Steps

1. Storyboards used for planning purposes follow the steps used in developing Affinity and Tree Diagrams (i.e., generating ideas and organizing tasks). For Storyboarding, the project title is followed by successive layers of detailed activities (i.e., acts and then detailed scenes within each act) that are grouped under header and subheader categories.

2. Storyboards that communicate the results of quality improvement (see the following example) help to document team activities. These Storyboards are often reduced to two- to six-page summaries using thumbnail graphics.

Example

The Check-Out Process Improvement Team is presenting the results of its project to the Board of Directors of the Vinyl Corporation. It is six months after the team began implementation of its recommended changes. Mr. Fetish wants to demonstrate the exciting results of this team and the entire Quality Management effort. The following Storyboard summarizes project findings.

Project: Video One Check-Out Process Improvement

This project analyzes the reasons for the long check-out time, which is the primary source of customer complaints. It is anticipated that improvements in the process will improve customer satisfaction and have a favorable effect on profitability.

Quality Improvement Team

Mr. Fetish	President, Vinyl Corporation
Geraldo	Store Manager, Video One
Ernestine	Cashier, Video One
Willard	Inventory Clerk, Video One
Ivan	Accountant, Video One

Step 1: Select Improvement Opportunity

After Brainstorming potential improvement opportunities, the team used List Reduction and then Criteria Rating to reach consensus on its first quality improvement project. The team developed the following Problem/Opportunity Statement.

Problem/Opportunity Statement

"As Is" Condition: Check-out time averaging nine minutes is the primary reason for customer complaints.

"Desired State": Reduce the actual check-out time by 50% in six months and reach benchmark standard in one year.

Note that the team modified its original statement after it collected preliminary data during Step 2: Analyze Current Situation.

Step 2: Analyze Current Situation

The team confirmed the existence of the perceived problem by conducting an in-store Survey of customer complaints and suggestions. They received 94 complaints, 48% of which were about the long check-out time (Figure 11.5).

The team also validated the check-out time by conducting a random sampling of times that averaged eight minutes (Figure 11.6).

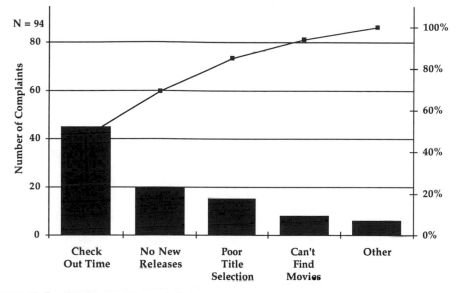

FIGURE 11.5   Customer Complaints

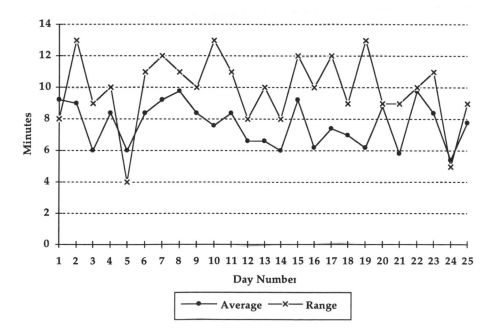

FIGURE 11.6   Average and Range of Check-Out Times

A Deployment Flowchart of the check-out process prepared by the team indicates that many steps in the process do not add value for the customer (i.e., steps with a theoretical time of "0" when no value is added). This flowchart revealed several problems that contribute to the long check-out time:

- Manual inventory records and invoices
- Separate storage area for rental tapes
- Manual credit card transactions

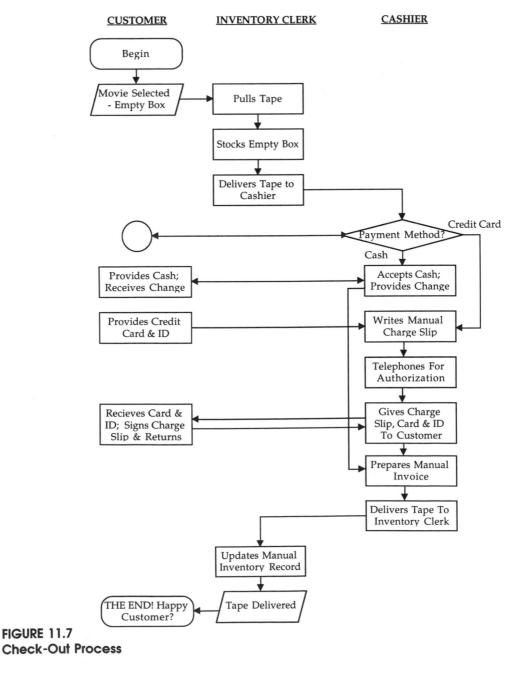

FIGURE 11.7
Check-Out Process

Step 3: Identify Root Causes

The team brainstormed potential root causes of customer complaints using the Five Whys technique. Several causes that affect check-out time were explored, including the manual activities noted earlier. The team's analysis also indicates that high employee turnover, possibly due to low pay, may cause some of the long check-out delays.

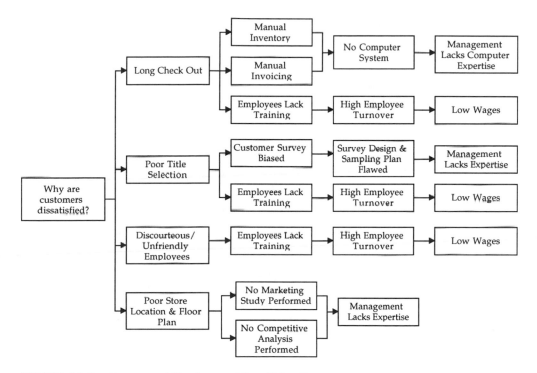

FIGURE 11.8   Causes of Customer Dissatisfaction

Step 4: Select and Plan Solution

The solutions recommended were planned with a phased implementation in the following sequence:

1. Remodel store and place rental tapes on display shelves:
   - Install security system
   - Expand display area by eliminating tape storage room
2. Automate credit card authorization
3. Automate inventory and invoicing activities
4. Retrain Inventory Clerk to perform market research duties

Step 5: Implement Pilot Solution

All recommended solutions, except the installation of the computer system for inventory and invoicing, were implemented without pilot activities. A one-week parallel of the new computer system was performed with the old manual system.

Step 6: Monitor Results and Evaluate Solution

The team prepared the following Flowchart to reflect the changes implemented. Check-out time has been reduced to 2.5 minutes, which reflects a 68% reduction in time. Rental volumes and profits increased 25% in the last six months.

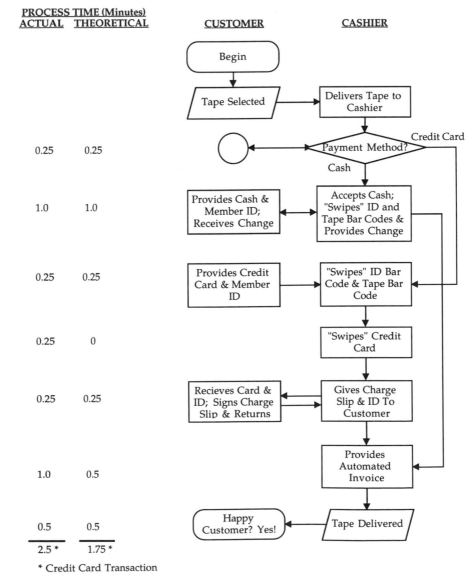

FIGURE 11.9
Revised Check-Out Process

TREE DIAGRAM

Tree Diagrams are described in Chapter Seven: Process Definition. In this first application during quality improvement, Tree Diagrams help to describe the sequential flow of a process in which several layers of detail join to form a single output. Next, Tree Diagrams were used in the search for cause-and-effect relationships using the Five Whys technique described in Chapter Nine: Analyzing Cause and Effect.

Finally, Tree Diagrams appear as a planning tool, which is the primary application of this powerful tool. In this application, the Tree Diagram resembles the planning Storyboard described in the previous section. The difference between the two tools is in name only.

CHAPTER TWELVE

MEETING MANAGEMENT TOOLS

Two meeting management tools are described in this chapter: Agendas/Meeting Minutes and Plus/Delta (+/Δ) Evaluations. Along with Storyboards, which were discussed in Chapter Eleven, these tools are valuable in documenting both the content and process employed by quality teams. They not only help provide a road map for others, but also provide a means to communicate team progress.

Often team effectiveness depends on how well the team manages its time during meetings. Agendas/Meeting Minutes help teams to manage both the process and content of meetings. Minutes document the discussions, decisions, and actions items of each meeting, while agendas ensure that future meetings are properly planned. Plus/Delta (+/Δ) Evaluations assist teams in improving both the content and process of a meeting. This critiquing can significantly improve team effectiveness.

AGENDAS/MEETING MINUTES

Description

Quality improvement depends on teams working together productively. In many cases, teams members are working together for the first time, and the team goes through stages of development (often categorized as the forming, storming, norming, performing, and adjourning stages). Although group dynamics is beyond the scope of this book, one tool assists work unit teams to effectively manage both the content and the process of meetings: Agenda/Meeting Minutes. This tool not only documents the discussions, decisions, and actions items of each meeting, but is also used to set the agenda for the next meeting.

In addition, the logistics of meetings (i.e., date, time, and location), attendance, and team roles are identified. Each segment of the meeting is defined as either information sharing or information processing (i.e., making decisions and assigning action items). Key elements of the next meeting agenda include the purpose of the meeting, desired outcomes, and the scheduled time for each topic.

Key Points

- Meeting planning consists of five key steps:
  - o Determine purpose and desired outcomes
  - o Decide if meeting is required
  - o Identify participants
  - o Assess team ability and authority relative to desired outcome
  - o Prepare agenda items and sequence
- Meetings generally follow a standard format:
  - o Review advance meeting agenda and revise as needed
  - o Review meeting minutes from prior meeting
  - o Discuss old business, including action item status
  - o Address new business
  - o Begin with information sharing, followed by information processing
  - o Summarize progress and new action items
  - o Critique meeting process and effectiveness
  - o Decide on agenda for next meeting: purpose, outcomes, and topics
- Agendas and meeting minutes become the road map to document team progress and keep the team focused on project content and the meeting process
- Information processing usually follows information sharing during meetings
- The minutes-taker assumes responsibility for preparing meeting agendas and minutes
- Meeting minutes and next meeting agenda should be reviewed with the team leader before distribution
- Agendas and meeting minutes should be maintained in a summary form and be limited to two pages total

Example

Meeting Agenda

PROJECT:

Date:	Time:	Location:
Leader:	Scribe:	Timekeeper:

Purpose

Desired Outcomes

Topic	Responsibility	Outcome	Time (min.)
Review agenda/outcomes	Leader	Revised agenda/outcomes	5
Status team progress/prior meeting minutes/action items	Leader	Consensus on team status	10
	All		
	All		
Break			10
	All		
	All		
Meeting summary	Leader	Decision/actions item review	10
Next meeting agenda	Leader	Next meeting agenda	5
Evaluate meeting	Leader	Plus/delta chart	5

FIGURE 12.1   Meeting Agenda

Meeting Minutes

PROJECT:					

Date: _____   Time: _____   Location: _____

Team Member	Attended	Role	Team Member	Attended	Role

Information Sharing: Main Points

Information Processing: Key Decisions

Information Processing : Key Action Items/Responsibilities

FIGURE 12.2   Meeting Minutes

Steps

Although the forms are self-explanatory, the following comments highlight their use:

1. Meeting agendas should always provide sufficient notice for participants to complete required prework.

2. Agendas should always be reviewed and revised, if needed, at the beginning of the meeting. During the meeting, times allocated for important topics should be revised, if needed, by team consensus. A new agenda for the next meeting should be developed at the end of each meeting.

3. Review the status of action items from prior meetings at the beginning of each meeting so that due dates are refreshed each meeting.

4. After the review of existing action items, discuss topics in order of importance. Information sharing (e.g., presentations) should always precede information processing (i.e., discussions, questions, and consensus decisions).

5. Minutes should include logistics, attendance, and meeting roles in order to document resource requirements.

6. Minutes summarize the main points of the meeting (from information sharing), key decisions (from information processing), and key action items and responsibilities (from information processing).

7. Meeting minutes should be completed as soon after the meeting as possible (within a day or two).

PLUS/DELTA (+/Δ) EVALUATION

Description

Plus/Delta (+/Δ) Evaluation is a tool that is used to critique meetings or team performance. It identifies things that worked well and areas that need improvement. Plus/Delta (+/Δ) Evaluations stress the need for continuous improvement in process (e.g., effective use of quality tools) in addition to improvement in content (e.g., amount of data).

Plus/Delta (+/Δ) Evaluation is similar to Balance Sheets (Chapter Six: Consensus) in that it is used to evaluate the pros and cons of alternatives.

Key Points

- Generates a quick list of what worked and identifies improvement opportunities for future meetings or activities

- Provides direct feedback to leader, scribe, and other team members
- Allows participants to critique and improve performance

Typical Applications

- Used to critique meetings or team performance in all types of activities
- Used at the end of a meeting or activity to improve the process of the next meeting or activity

Example

MEETING CRITIQUE		
Plus (+)	**Delta (Δ)**	**Recommendation**
Good teamwork	Late start	Change meeting to better time
Ended on time	Agenda not finished	Develop realistic agenda
Reached consensus	Reached consensus too fast	Review group-think concepts
Great ideas	Too many anecdotes	Limit war stories
Had fun	Breaks too long	Take only 10-minute breaks

FIGURE 12.3   Plus/Delta (+/Δ) Evaluation of Team Meeting

Steps

1. Identify the subject of the evaluation, such as meeting, project, or specific activity.

2. Set up three columns on a flip chart. Put a plus (+) above the left column, a delta (Δ) above the center column, and "Recommendations" above the right column. Underline column headings and draw a vertical line between the symbols down the length of the flip chart.

3. Brainstorm both positive areas and areas needing improvement. Some items might be both a plus and a delta, depending on the circumstances.

5. Discuss each delta to determine if there is consensus that improvement is needed.

6. Make recommendations for improving recurring or significant deltas.

CHAPTER THIRTEEN

BENCHMARKING

Benchmarking is an externally focused quality improvement technique that seeks "best" practices in other organizations which become partners in the study. These partners range from other divisions in the same company, to direct competitors, to competitors outside the industry that perform similar functions. After identifying superior performance and understanding the drivers of that performance, the partners can apply "best" practices to their own organizations. Because it is unlikely that any one organization does everything well and will be unable to learn from others, all partners can potentially gain from Benchmarking.

A summary of the Benchmarking technique is presented in this chapter. Teams that need more detailed information should consult *Benchmarking, The Search for Industry Best Practices that Lead to Superior Performance* by Robert C. Camp or *Benchmarking: Focus on World-Class Practices* by AT&T (see Appendix B: References and Resources).

BENCHMARKING: THE SEARCH FOR "BEST" PRACTICES*

Introduction

Benchmarking is not a new concept. It shares many elements with a multitude of activities, ranging from competitive analysis and Total Quality Management to ancient warfare. Since its inception as a formalized process at Xerox in 1979, it has

---

* This chapter was originally written by the author for publication as the Benchmarking chapter in the Society of Manufacturing Engineers' *Continuous Improvement Handbook*. It was entitled "Benchmarking: The Search for 'Best' Practices" and is reprinted with permission.

evolved into a technique used by the majority of Fortune 500 companies. Not all companies that try it are pleased with their success, but many are. Some companies have embraced the practice and want to share their experiences by offering training to others. Notable amongst this group are AT&T, IBM, and Motorola. What began as a process to improve the design and manufacture of copiers is now used in a variety of industries and functions. Today, this diversity of applications has gone well beyond manufacturing industries and now includes studies in government, education, agriculture, health care, and financial institutions.

The purpose of this chapter is to answer some of your questions about Benchmarking and to help you understand why so much attention is being focused on the process. More importantly, it provides guidelines to help teams get started and describes pitfalls to avoid for teams that have begun the journey.

Wake-Up Call for Xerox

With patent protection for the xerographic process, Xerox had prospered with little or no serious competition. Unfortunately, they had created a large, bureaucratic, complacent organization. The expiration of patent protection was Xerox's wake-up call. In the late 1970s, Xerox's domestic market share declined to 22% (unit shipments) and their revenues declined from 82% to 42% of the copier market. Their low-volume copier market was essentially lost (1% share) to the Japanese, who were selling copiers through wholesale dealers like 47th Photo in New York for prices below Xerox manufacturing costs.

Xerox met the challenge in 1979 with a strategic action called Competitive Benchmarking, which focused on Japanese design and manufacturing techniques, particularly those of its joint venture partner Fuji-Xerox. The quality transformation at Xerox began in 1983 and Benchmarking was one of three key processes. The change at Xerox was dramatic: in 1986 Xerox launched a family of mid-volume copiers that were designed in half the time and cost essentially half as much to design and manufacture as the previous generation introduced four years earlier. In 1987, a DataQuest newsletter stated that the "Benchmarking program deserves the lion's share of the company's turnaround in recent years."

During the 1980s, Xerox applied Benchmarking to many functional areas of the company. A 1982 study with L.L. Bean, which focused on logistic and distribution issues, proved that Benchmarking of functional areas in unrelated industries could be beneficial. Xerox enjoyed a 10% improvement in logistics and distribution productivity in the early 1980s, with 30% to 50% of the gain attributed to the L.L. Bean study.

By the end of the decade, Xerox's low-volume market share grew to 20% in the United States, with product produced in both the United States and Japan. The quality transformation at Xerox paid off and Benchmarking played a key role in that rebirth.

Benchmarking and Competitive Analysis

Benchmarking can be confused with competitive analysis, which is a related but different exercise. Benchmarking shares many elements with competitive analysis. It often seeks information in the public domain and it looks for creative ways to obtain and analyze data. There are significant differences, however. Two key distinguishing characteristics of Benchmarking are that the organization being studied is cooperating as a partner in the study and the focus of the study is on processes and practices, not just performance measures. Table 13.1 summarizes the major differences for organizations looking at the external environment.

TABLE 13.1   Characteristics of Competitive Analysis versus Benchmarking

Characteristic	Competitive Analysis	Benchmarking
Approach	Independent	Partner
Performed By	Individual	Team
Target	Competitor	"Best" Practices
Focus	Performance Measures	Processes and Practices
Objective	Competitive Intelligence	Process Improvements

The current literature on Benchmarking offers a wide array of "Benchmarking" categories. Many of these distinctions are quite arbitrary and have nothing to do with the process itself but deal more with the partners chosen and area of the business being analyzed. The Benchmarking process does not vary significantly with the three categories of partners selected: internal organizations, direct competitors, or non-competitors. This chapter assumes that there are two forms of Benchmarking: strategic and functional or operational. We will address the latter form of Benchmarking, which focuses on the operational processes and practices and services offered by an organization. Strategic Benchmarking focuses on strategic marketing, financial, organizational, and technological issues facing an organization. Although it is important to benchmark how other organizations select and deploy the vision, goals, and policies to work units (i.e., organizational goal alignment/policy deployment), it is not the subject of this chapter.

Table13.2 summarizes Benchmarking categories using this revised definition.

TABLE 13.2   Revised Benchmarking Categories

Common Categories	Revised Definition
Strategic Benchmarking	Strategic Benchmarking
Process, Service, and Functional Benchmarking	Functional/Operational Benchmarking
Performance, Competitive, and Industry Benchmarking	Industry Analysis

Industry analyses performed by professional and trade associations often have the look and feel of cooperative, competitive analysis, but on a broader scale. Often, they go beyond the mere capture of performance measures and include outcomes (i.e., measures of business or process effectiveness and customer satisfaction) to help identify "best" practice companies. Generally, these studies include organizations of various sizes in a variety of industries, which provides desirable diversity, but the identity of the participants is usually protected. This seriously handicaps the studies for use in functional/operational Benchmarking, however. These studies do provide valuable information on data not normally available in the public domain, and when these studies are repeated, they provide important trend information.

Benefits of Benchmarking

The United States is no longer competitive in many global industries. As with other quality improvement efforts, improved methodologies result in higher productivity and lower costs. This can only occur if organizations understand the need to change, are willing to change, and have an idea of the outcome after changes. Benchmarking is particularly helpful in validating proposals for change. Although circumstances are different in other industries, Benchmarking often results in creative imitation and the adoption of new practices that overcome previous industry barriers. This search for diversity and for innovative breakthroughs applied elsewhere is at the core of Benchmarking benefits. By sharing information, all parties benefit, because it is difficult to excel in all activities. Sharing information and data is often the first hurdle to be overcome in the Benchmarking process. Do not, however, attempt Benchmarking in areas in which trade secrets or sensitive information determines the outcome of the process.

The benefits of Benchmarking, as exemplified by Xerox, are fairly well documented by many organizations in a wide variety of industries. Benchmarking, used in conjunction with other quality techniques or used alone, can influence how an organization operates. If the search for "best," or just "better," practices is performed correctly, then the likelihood of successful outcomes is quite high. This, however, assumes that pitfalls are avoided and prerequisites have been met before Benchmarking is initiated.

Prerequisites for Success

Management commitment and support can overcome many of the barriers to successful Benchmarking. The key requisite for success is the organization's readiness to accept change, given that it comprehends the need for change. Rather than resting on their laurels and previous success, organizations must become receptive to new ideas. Additionally, sufficient resources need to be allocated, and both awareness and skills training need to be available.

Benchmarking shares success factors with other quality management processes. Teamwork, analysis of data, decisions based on facts, focus on processes, and continuous improvements are shared characteristics. The need for leadership, a customer focus, and empowered employees are equally important. For many organizations, success in quality management or Benchmarking requires that reorganizations and culture barriers (such as NIH ["not invented here], which exists in may professional disciplines) be addressed before beginning the journey.

As in other quality management activities, there needs to be someone to manage the introduction and application of Benchmarking in the organization. The tasks to help internalize the Benchmarking process within the organization include a wide variety of actions, such as:

- Identifying a Benchmarking champion

- Cascading just-in-time skills training to create centers of competency

- Coordinating projects and interaction with consultants

- Formalizing gateways to outside organizations

- Establishing networks, information databases, and newsletters

Not all of these tasks need to be in place before the first Benchmarking project, but they should receive consideration in implementation planning.

A final prelude to Benchmarking is often referred to as "step zero." In this preliminary step, a strategic and competitive assessment is performed to establish primary goals, objectives, and performance measures of success. The strategic assessment leads to the organization's vision of the future. This vision then gets translated into organizational goals, policies, and, finally, operating objectives that are both measurable and actionable. Benchmarking teams need to have some sense for the strategic direction of the organization so that their efforts are aligned with the organizational goals.

Additionally, Benchmarking teams need a clear understanding of both internal and external customer needs and expectations. Without this, they will have difficulty selecting important subjects first since they lack knowledge of critical success factors necessary to satisfy those customer expectations. An understanding of competitive strengths and weaknesses provides additional background that aids the selection process. Strategic and competitive assessments not only tell the team what is important to focus on, but also give them an idea of how effectively they are currently achieving important goals. Finally in step zero, documentation of key processes (e.g., Flowcharts and baseline performance indicators) and practices finishes the desirable preconditions before starting.

An example will be used to illustrate the key points in this chapter: we will assume the case of a firm that designs, manufactures, sells, and services industrial/commercial equipment. Any number of products could be used, but we will assume that equipment is HVAC (heating-ventilation-air conditioning) sold primarily in North America. The firm specializes in equipment for small office and industrial buildings, and we will assume that it has only two direct competitors in its market niche. The firm, Clear Air, Inc., has just completed a strategic and competitive assessment. It discovered that its customers are very satisfied with low-cost new products that can be designed and manufactured with very short lead times. Clear Air assesses its critical success factors to be innovative new products; fast response for design, manufacture, and installation; and low-cost reliable products.

Eight-Step Benchmarking Process

The Benchmarking process consists of three general activities: planning, analysis, and integration/action (see Figure 13.1). Overall, the process follows the Plan-Do-Study-Act cycle of all quality processes. Within each step, this logical cycle is present at a lower level of detail. We will use an eight-step process (below) that can be easily linked to other Benchmarking processes (such as the Xerox ten step, the Alcoa six step, and the AT&T nine step). This is important because you will have Benchmarking partners that will likely select one of these models and you need a common language for effective communication. Each step discusses the numerous pitfalls that might be encountered. Benchmarking is like other quality processes in that learning is often enhanced by the failures along the way.

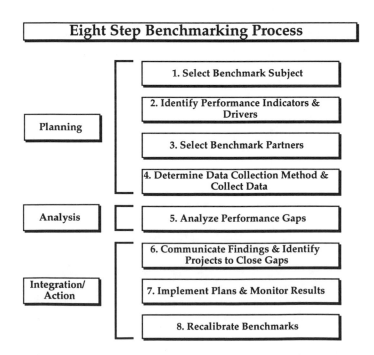

FIGURE 13.1   Eight-Step Benchmarking Process

Planning

Selecting the Subject

The first step in the planning phase is to select a subject. Quality Steering Committees, senior management, functional department heads, or quality teams generally perform this function. The first Benchmarking project may not be the most critical problem, but it may serve an important educational purpose. Additionally, it may help to convince management if they are skeptical of the value of Benchmarking. Subsequent projects should be ones that are critical to the success of the organization. Key questions that can help to focus on the subject are:

- Where are competitive pressures and current problems occurring?

- Where is the most time and money spent?

- Where are the production or design bottlenecks?

- What areas are critical to the success of the organization?

- Where can the greatest improvement occur?

The selection of a Benchmarking subject is similar to the methodology used to select processes for improvement by internal quality teams. The primary difference is the external focus of Benchmarking and the added cost, which limits the processes addressed. Because organizational outputs generally depend on cross-functional processes and the practices associated with them, the selection task depends on identifying key processes and deciding which ones are the most important, which ones need improvement, and which ones will yield the greatest return if improved.

One technique that is quite useful in this ranking is the Prioritization Matrix (generally the analytical hierarchy process) coupled with assessments of process effectiveness. The disadvantage of this technique is that it requires significant effort by those with subject matter expertise. The primary advantage is that it is done only once (or easily repeated if necessary) and provides a prioritized order of all major processes needing improvement. This approach is often used by senior management and Steering Committees, for example, when a more global view of the organization is necessary. Its strength lies in the focus on outcomes or goals, the identification of success factors needed to reach the goals, and the identification of key processes that drive performance.

A less rigorous approach can also be used, which relies on the typical problem-solving/process improvement process: identify potential problem areas/process improvement opportunities, identify customers and outputs, conduct preliminary analysis of potential causes, and select key causes. Both of these techniques help to steer teams past the first serious pitfall of Benchmarking: attempting several difficult projects simultaneously.

Another common tool used to prioritize projects and decide which ones to pursue is the cost–benefit analysis. Here, the potential benefits from a changed process or practice are compared with the cost of the study and costs to implement potential recommendations. Estimates of the cost of performing a Benchmarking study, however, vary significantly depending on the scope, logistics, and maturity of the teams. Typical projects (five to ten partners, performed by a mature team) might require six team members working 25% of the time for six to ten months, or three-quarters to one-and-a-quarter man-years.

One final note on selecting a subject: avoid selecting subjects that involve processes or activities that involve trade secrets or sensitive information. Even though you might stipulate off-limit areas, the risk of divulging sensitive information may not be one that your management will be willing to take.

Forming the Team and the Project Road Map

Next, a team needs to be formed (or modified if it is an existing quality team) and a project road map developed. These are integral activities in the subject selection step. The team should include four to eight members who are subject matter experts and represent the various functions affected by the project. Teams composed of staff personnel who are not owners or stakeholders of the process being studied often results in recommendations that meet significant resistance during implementation. This pitfall can easily be avoided by including members from all functions affected.

Teams often include several people in an advisory capacity: the project sponsor or "customer" and a Benchmarking "coach" (an internal or external consultant). This is a way of keeping the customer informed of the team's progress and the findings. Both the customer and coach should provide advice rather than dictate content.

The road map serves several purposes: it keeps the team on track and it documents activities and decisions. This helps to educate others and to document the team process for later use. Another advantage is that clearly defined projects help teams to ensure that the project scope is achievable. Basic elements of the road map include:

- Project scope and objectives

- List of activities

- List of deliverables

- Individual roles and responsibilities

- Meeting and review schedules

- Issues

During the last year, Clear Air has lost several contracts to a competitor that was able to deliver against a very tight delivery schedule. In the past, this competitor was unable to perform that quickly. Examples of projects that might be considered are manufacturing cycle times and product development cycles. The team might want to explore the impact of reducing the work-in-process inventory with the application of just-in-time inventory management and the impact of shortening the design cycle with concurrent engineering techniques. These preliminary subjects are quite broad and the team will use Top-Down and then Deployment Flowcharts of key processes to determine non-value steps, sequential flows that can be parallel, and sources of process variation. Next, the team will collect additional data to confirm problem areas with a potential for significant improvement. They will then be able to do a preliminary cost–benefit assessment to help prioritize and select Benchmarking subjects. An alternate approach would be to use a Prioritization

Matrix with assessments of current levels of process effectiveness to gain a more global perspective.

Performance Indicators and Drivers

This step begins with the documentation of processes and practices associated with the subject, if not already completed. The elements of a process will be assumed to include customers, inputs, outputs, activities, decisions, sequence, and responsibilities for adding value. Practices can be process specific or have an impact on multiple processes and generally define how activities and decisions are performed. For example, product assembly can be defined by manual or mechanical practices and communication can be either verbal or written. Documentation activities include the development of process Flowcharts and the collection of baseline data from internal operations as well as relevant public domain sources (e.g., trade and professional association studies, newsletters, and trade publications). The primary goal is to identify the vital few performance indicators that confirm superior performance and to identify those processes and practices that drive performance. This search for cause and effect will be followed by the identification and documentation of internal process variables and attributes. A table of data and information that is both quantitative and qualitative (Figure 13.2) facilitates the development and analysis of indicators and drivers. It will also be used to compare inputs from Benchmarking partners.

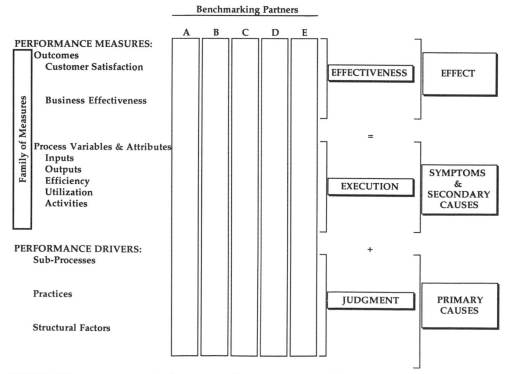

FIGURE 13.2 Table of Performance Indicators and Drivers

The key to finding the vital few measures starts with the identification of customer satisfaction measures and measures of business effectiveness. Finding this measurement of results, or outcomes, is a difficult yet necessary task in successful Benchmarking. The difficulty arises because effectiveness depends on selecting the "right things to do" and "doing things right," or judgment and execution. Judgment creates the system of performance drivers (i.e., processes, practices, and structural factors), while execution determines how productively the system performs (i.e., efficiency times utilization). Poor results usually occur when either or both of these elements fail. Superior execution can seldom overcome wrong or poorly chosen drivers, however. The search for key drivers (i.e., causes) is simplified if the failure can be identified as either one of judgment or execution.

This dilemma will plague us in selecting the vital few measures, in linking these measures with performance drivers, and later on, in selecting partners and analyzing data. As would be expected, unfavorable outcomes often mask excellent processes that are executed poorly. This is the "halo" effect, where an organization that performs well is often assumed to do everything well, and conversely, an organization that performs poorly is assumed to do everything poorly. In Benchmarking, organizations with poor outcomes are generally ignored because of credibility issues.

It is important to select performance indicators that cannot be manipulated or result in suboptimization (win the battle, but lose the war). Often the best measure may be one that is not currently in use or is felt to be too difficult to measure on a continuous basis. The ideal is to pick ratios that can be continuously measured, but a good measure monitored periodically is better than a poor one measured continuously. A cost accounting labor overhead rate that is monitored regularly is a good example of a poor indicator, because there are so many factors that influence the ratio. Because accounting systems vary significantly, it is often difficult to use financial data without major adjustments to ensure comparability. In most cases, measures of time (e.g., machine-hours, man-hours, and elapsed time) and various dimensions of quality (such as failure rates, acceptance levels, defects, and yields) are better than financial measures.

Indicators are only a means to an end rather than an end in themselves. They track change that results from the implementation of improved processes and practices and are expected to improve over time as process variation is reduced and the process velocity is improved (e.g., by eliminating non-value steps or changing work and information flows). (Process Flowcharts are particularly useful in identifying where process variation occurs, where failure data need to be collected, where value is added, and where parallel flows can replace sequential flows.)

After identifying the vital few measures, the team needs to understand the causal link between the performance drivers and these indicators. This critical step requires subject matter expertise using various techniques such as fishbone diagrams and Interrelationship Digraphs. The team should, however, remain open-minded regarding this relationship since the study might uncover other performance drivers and a better perspective of the cause-and-effect relationship.

During the discovery of these relationships, the team should not lose sight of the customer, customer expectations, and success factors defined by the customer.

Finally, the team needs to address a special category of performance drivers: structural factors. These are the culture, organization structure, technology, environment, and any other administered factor (such as contractual agreements) that might have an influence on the performance indicator. Structural factors create barriers that must either be corrected or accepted as a given but should never be ignored. Acknowledging these barriers helps to reconcile differences in performance between Benchmarking partners during the analysis stage.

The team would then define the performance indicators and some preliminary performance drivers, depending on the specific subject selected. The outcome measures of short manufacturing lead time and short design cycle time will be used as measures of business effectiveness with an assumed direct relationship to customer satisfaction. In addressing the manufacturing lead time problem, the team might want to look at several indicators, such as work-in-process inventory turnover, rework costs as a percentage of unit manufacturing costs, number of suppliers, incoming raw material quality, final inspection yields, number of quality defects, raw material lead times, and supplier delivery frequency. An exploration of the drivers of manufacturing cycle times might include a review of some of the following: material requirements planning (MRP) systems and practices, MRP software, purchasing policies and practices, discrepant material handling, supplier alliances, supplier quality certification process, and order entry and factory scheduling systems and practices.

On the design side, the appropriate indicators might include new product design cycle time, variant product (i.e., second generation) cycle time, engineering change order frequency, unique drawings per engineer, approvals per drawing, organization layers in engineering, and percentage of drawings using computer-aided design (CAD). Performance drivers in the design area would likely include: extent of CAD, product development process and practices (including customer Focus Groups, early supplier involvement, pilot production, and manufacturing involvement), engineering approval and review process, and robust design practices.

The primary focus should be on those areas where major variations in performance occur, process steps do not add value, and work flows are not optimized.

Selecting Partners

The selection of partners often involves the use of external data and information sources. The secondary research performed while establishing baseline information and data will often uncover articles and sources helpful in identifying organizations that perform well in your subject area. If these sources do not uncover potential partners, continue your review of periodical and trade sources and begin a search focused on direct contact with trade associations, consultants, customers, suppliers, employees, Baldrige Award companies, and Benchmarking clearinghouses.

Most organizations performing Benchmarking designate a "champion" to facilitate training and project management and to provide a gateway for initiating projects. These champions form an informal network (sometimes formalized by their joint membership in various associations) that can be instrumental in locating partners. Tapping into the network is like finding the mother lode ore deposit.

Partners fall into two general categories: other internal units and external organizations. If there are other units within your organization that are performing similar functions, they should always be included. External organizations include:

- Direct competitors

- Industry peers that serve a different market

- Companies that serve your market in a related but different industry

- Organizations outside your industry that perform similar functions

The first category is often ignored because of the fear of divulging sensitive information, but competitors are often willing to participate in non-sensitive areas or when the anonymity of participants is protected by an outside third party, such as consultants or trade associations. The last two categories are often the source of greatest innovation because they are not generally influenced by the paradigms of your industry.

Once a preliminary list has been developed, partners are selected using the criteria below. Here, the selection process will be based on data and information available in the public domain. Because such information is often general and not specific to the subject selected, it is important to be careful of the "halo" effect.

- Profitability

- Industry stature and potential for "best" practices

- Functional and process expertise and potential for "best" practices

- Ease of obtaining information (indicated by prior participation in earlier studies)

- Relevancy

- Potential for innovation

The benefit of selecting organizations outside of your industry can be lost if issues of relevancy and credibility of the organization are not addressed in the selection process. The relevancy issue should be discussed immediately with partners that agree to participate in order to avoid wasting each other's time. A brief discussion of the subject is generally sufficient to help both parties conclude that they have the potential to learn and apply innovations from each other. The search for "the best" organization is generally a waste of time because of the law of diminishing return. Remember, the objective is continuous improvement, and future studies, if warranted, might find "the best." Finally, do not place the ease of collecting data above the potential for "best" practices in the selection criteria weighting.

Most Benchmarking studies include five to ten partners. Exceeding ten partners complicates project management and should be avoided in most studies. (The primary exception is industry studies funded by a trade association or by participants in the association study. In these studies, the scope and time frame are extended beyond those recommended for functional/operational Benchmarking within an organization.)

To obtain the participation of partners, you must be willing to share information and data about your own organization and provide a copy of the final report. In many cases, receipt of the final report is sufficient inducement to participate. The benefits of Benchmarking should not be oversold, however.

The search for partners in our example would use the same sources discussed above. Trade publications and organizations would be those that address HVAC design, manufacturing, and marketing. In addition, trade sources serving the same customers (e.g., elevator manufacturers serving small office and industrial building contractors) or related industries (such as residential HVAC and major appliance manufacturers) could be explored. Given the conditions described above, Clean Air's direct competitors should be excluded, but industry peers serving other markets (i.e., companies producing HVAC equipment outside their current geographic market or producing different HVAC products and customers) should be considered.

Data Collection

Data collection can be the most difficult step in the Benchmarking process. The primary objective is to gather information and data to confirm superior performance and to uncover "best" practices without burdening the partners with long, time-consuming data collection methods. Although data and information are collected in establishing baseline measures and in selecting partners, the majority of the data for most projects are collected in this step.

The types of data and information required often define the method for obtaining them. For most Benchmarking projects, primary research is required because the level of detail goes far beyond the level available in secondary published sources. The type of primary research generally involves a Questionnaire or Survey document, which will be used in combination with other activities. These documents address both the quantitative performance indicators ("what") and the qualitative performance drivers ("how to's"). Most organizations prefer to collect all information with one document, with follow-up questions to validate the data; other teams prefer two documents, the first to collect the quantitative data and the second for the qualitative information.

The typical sequence of data collection begins with a Questionnaire mailed to each partner, which is followed by a telephone interview to clarify key points after the written response is received. An alternate and commonly used approach is to fax the document and then conduct a telephone interview. This works well for short Questionnaires that require minimal data collection. Be sure to schedule all telephone interviews in advance. Facility visits and meetings to discuss preliminary findings generally follow the preliminary data collection activity. It is recommended that two or three members of the Benchmarking team conduct visits and that they adhere to the scheduled time and agenda. Each member should have a clearly defined role in the visit (such as note taking, asking questions, process guidance, etc.), and the team should conduct a debrief as soon after the visit as possible (i.e., less than two days). Most projects include a final meeting of partners during which the study findings are reviewed and discussed.

The actual methodology depends on data and information requirements (amount and accuracy of data) and other criteria such as costs, project schedule, and logistics issues. Facility visits and meetings to discuss preliminary findings are not a prerequisite for successful Benchmarking, however. Visits to partners in other industries are generally recommended because the need to discuss normalization of industry differences often requires more time and interaction than is possible over the telephone. Visits to partners who will help to pretest the Questionnaire are also recommended, but trips that add little value to the process (also known as "industrial tourism") should be avoided. Be prepared to answer questions about your organization and its answers to the Questionnaire, but make sure your responses do not influence your partner's answers (i.e., by providing your answers first).

The first step in developing a Questionnaire or Survey is to list all questions that need to be answered in order to validate the existence of superior performance. These questions are followed by ones that ensure that everyone is measuring the same thing and that a common denominator can be identified. The data capture mechanism (Questionnaire, Survey form, computer input form, etc.) should be logical and easy to use. Limit the data captured to the vital few indicators, which not only minimizes the respondent's time, but also facilitates data input and analysis. Detailed, statistically valid data are mandatory for some analyses, but a lower level of detail is often acceptable for most Benchmarking projects. Open-ended questions about relevant processes and practices and requests for process Flowcharts usually follow to uncover the performance drivers.

Next, the team should answer all questions and, if possible, pretest the Questionnaire with an internal unit and one external partner. Questions that are difficult to answer or not essential must be revised or eliminated. Add appropriate questions suggested during the pretest. Perhaps the most difficult task in developing a Questionnaire is to ensure that it is focused on a clearly defined subject and scope. This will help to avoid long, detailed Questionnaires that potential partners will either ignore or provide incomplete answers to.

Because Benchmarking requires interaction with other organizations, ethical and legal issues need to be addressed. Always consult legal counsel when in doubt. Exchanging cost and marketing information, for example, can be construed as an action leading to restraint of trade, which must be avoided. Acquiring or disclosing trade secrets or sensitive information can likewise create legal problems which can be easily avoided. In many cases, an outside consultant who protects the anonymity of the participants of a study can help to avoid problems that result from face-to-face interaction between partners.

From an ethical standpoint, the key cautions are to not misrepresent yourself, to respect the confidentiality of information, and to use the information for its intended purpose. Results should not be shared with third parties until all partners have approved this disclosure. During the project, partners should be prepared for each activity, share data and information, honor commitments, and work through designated channels within partner organizations.

The initial stages of collecting performance data can rely on mailed Questionnaires and telephone follow-up. The Questionnaire needs to address the specific questions related to the subject plus more general demographic questions that are helpful in analyzing organizations not in the same industry. Requests for process Flowcharts and descriptions of various practices can also be handled in the same way. To interpret Flowcharts and to fully understand processes and practices, it is likely that the Questionnaires would be followed with plant tours and personal contact with partners. This is particularly important when partners are not in the same industry and a thorough understanding is necessary to normalize the data (i.e., for apples to apples comparisons) for analysis, which is the next step.

Analysis

The objective of the analysis step is to identify the "best" performing organization and to determine the reasons for the superior performance. Because the Table of Performance Indicators and Drivers (Figure 13.2) highlights the key elements needed to compare partners, it becomes a useful tool in analysis. The performance indicators define the benchmark standard and the gaps in performance for each participant. The processes and practices of the "best" organization are the benchmark performance drivers that each partner will try to creatively imitate.

Identifying this cause-and-effect relationship between performance drivers and the resultant measures is the most challenging part of Benchmarking. Here, the impact of structural factors on performance is an important consideration, because many of these factors relate to qualitative rather than quantitative information and are often more difficult to analyze. Analytical problems frequently occur when diverse organizations are being compared or when it is difficult to isolate organizational efficiency (i.e., "doing things right") from organization effectiveness (i.e., "doing the right things right"). In short, a good process that is executed poorly can easily be overlooked.

The primary concerns in data analysis are both the accuracy and validity of the data. Difficulties can arise if there is too much, too little, or inconsistent data; this often results in analysis paralysis—an endless search for trends or relationships. The desire for excess precision can lead to a similar trap. The purpose of keeping the scope of the study focused on critical success factors and the vital few performance indicators is to avoid these problems. Aside from the volume of data collected, the next most challenging task is to normalize the data from organizations that are not identical. This problem is particularly acute when financial data are used, particularly if the partners reflect the recommended diversity. The caution about using financial data becomes evident in the analysis step.

Because benchmark performance is a moving target, it is necessary to forecast performance into the future based on current trends for both your organization and the "best" organization. The Xerox "Z" chart (Figure 13.3) is particularly useful in graphically portraying the competitive gap for key indicators and the extent of improvement required to close the gap. This chart highlights the need

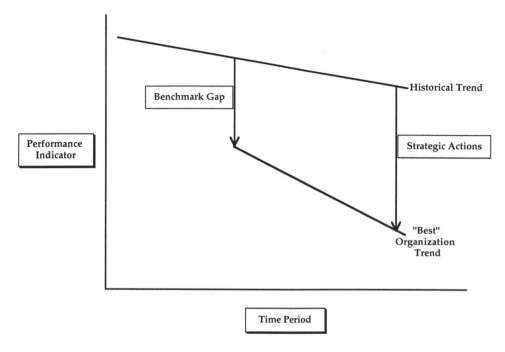

FIGURE 13.3   "Z" Chart

to take strategic actions as well as continue or improve the current level of process improvement.

Progress in achieving performance improvements is then monitored relative to the forecasted trends. Keep in mind that indicators are only a means to an end rather than an end in themselves. They track improvements that result from changes in performance drivers; they track change from both the historical trend and the strategic actions taken as a result of the Benchmarking study. The mentality should be to measure, compare, and improve the processes instead of mandate a fixed standard and hope that performance improves.

For most projects, it is advisable to distribute preliminary conclusions, along with supporting data, to participants to have them validate their data inputs and comment on the preliminary reasons for superior performance. Large, apparent performance gaps need to be treated with great sensitivity to ensure that partners have the opportunity to explain the differences and correct possibly erroneous data.

The Clear Air example does not require special discussion here and would be analyzed in the manner discussed above. The main differences between internally focused problem solving/process improvement and the Benchmarking process usually merge in the analysis stage, which assumes that data from different industries have been normalized for comparison purposes. The primary reason for conducting a Benchmarking study is to learn about and observe innovative processes and practices. Once this search is complete, the gaps defined, and reasons for the gaps understood, the organization can then creatively adapt these innovations in the next phase, where it takes strategic actions to change.

Integration and Action

The objectives of the integration and action steps are to:

- Obtain organization buy-in
- Initiate projects to close gaps
- Implement plans developed in these projects

Benchmarking is often used as a catalyst for change by adding credibility to proposals for change, by helping to break out of industry paradigms, and by creatively imitating the innovative "best" practices of others. Before the team attempts to obtain organization buy-in, it must organize and present its findings. The team must reach consensus about the findings and its recommendations for action.

The key to getting favorable responses to findings depends on how the team was developed (e.g., did it include stakeholders of the process) and how the project was managed (e.g., was management involved in a periodic formal or informal review process). Unfortunately, Benchmarking projects can easily become sidetracked in the integration and action steps. Resistance to change, the primary barrier, can be minimized if the process is managed properly and, more importantly, if the key prerequisite to success has been satisfied (i.e., management support and willingness to change).

If the team has focused on critical subjects, recommendations will probably require strategic actions and a new project to plan and implement the proposed change. Often the Benchmarking team is not well suited for both communicating findings and implementing plans for change. To prevent this, the team needs to be modified as it moves into this phase. The team needs to be committed to not only conducting the Benchmarking study, but to following the implementation of recommendations as well.

Since the recommended actions will likely have an impact on various strategic and tactical plans of the organization, the team must consider the integration of recommendations within the planning process. Finally, the recommendations need to address the responsibility for change, the timetable, and the estimated cost and benefits of the change. Realistic schedules and financial estimates are necessary to ensure the team's credibility.

Perhaps the most serious mishap in Benchmarking occurs when implementation is made in large, diverse organizations without pilot programs to test the impact. This caution is common to all change actions within an organization. Failure here should not be associated with the Benchmarking process, but rather with the change process within the organization. An effective sequence for pilot activities might begin with a small-scale pilot that is closely monitored. Other factors that have an impact on the process need to be isolated during the monitoring to ensure that changes result in the desired outcomes. Once success is obvious, results need to be communicated and, if needed, a training curriculum developed to facilitate the cascade of change. Training then becomes the vehicle for standardizing the change throughout the organization.

When the first project is completed satisfactorily, the organization is ready to tackle the next most critical subject. Before leaving the first project, the team needs to address one final issue: it needs to decide when it will repeat the Benchmarking of the new standard. Subsequent recalibrations are often triggered by a reasonable period of time (say, two to three years) or some specific change in performance indicators that are measured periodically.

Key Lessons and Next Steps

Benchmarking, like other quality techniques, works in organizations that have met certain prerequisites and that do not have major impediments to success. It is not a panacea, fad, or quick fix that is performed once, but rather a continuous process that is internalized within the organization. To be effective, Benchmarking works best in an environment in which teamwork and management support are evident. Teams need to focus on results as measured by performance indicators, but the organization must change the performance drivers to achieve benchmark standards. Target setting alone, without changes in the organization culture, other structural factors, processes, and practices, can only bring temporary relief. Table 13.3 lists the most common pitfalls experienced in Benchmarking.

TABLE 13.3 Common Pitfalls in Benchmarking

- Lack of management commitment and involvement
- Not applied to critical areas first
- Inadequate resources
- No line organization involvement (process owners and stakeholders; staff exercise only)
- Too many subjects; scope not well defined
- Too many performance measures
- Critical success factors and performance drivers not understood or identified
- Potential partners ignored: internal organizations, industry leaders, or friendly competitors
- Poorly designed Questionnaires
- Inappropriate data collection method
- Too much data; inconsistent data
- Analysis paralysis; excess precision
- Communication of findings without recommendations for projects to close gaps
- Management resistance to change
- No repeat Benchmarking

Benchmarking often starts in an organization because a few key people have educated themselves on the process and its benefits. Members of the first team generally have read books or articles on Benchmarking, have attended training seminars or workshops, and have problems that lend themselves to an externally focused problem-solving effort. If this team is successful and communicates its success within the organization, the technique will be applied by others.

The first team members form a Benchmarking competency center, and they are often relied on as coaches in subsequent Benchmarking studies. Their educational efforts will be enhanced when the organization designates a champion to promote Benchmarking and help manage the implementation of the technique within the organization. The management of Benchmarking (i.e., from the organizational level, not the project level) includes the cascade of just-in-time skills training, coordination of projects and contacts with partners, and the establishment of networks, newsletters, and project file databases.

For large organizations, the implementation process will take years; initial doubts will be eliminated as the organization continuously improves, as it strives to be the "best," and as it sets the standard for others to follow.

CHAPTER FOURTEEN

QUESTIONNAIRES

Questionnaires are useful in collecting data during quality improvement and are an integral part of the primary field research required to do Benchmarking. A well-designed Questionnaire can be invaluable in collecting valid, reliable, unbiased data. Questionnaires are usually completed by respondents in writing, but they can also be used as the basis for interviews (either in person or via telephone). Both Questionnaires and interviews are a form of Surveying in which only a sample of the population is included. If the entire population is involved, the Survey is a census.

This guideline was originally written to satisfy Benchmarking requirements, but it has been modified for inclusion here. Teams that need more detailed information should consult *The Survey Research Handbook* by Pamela L. Alreck and Robert B. Settle, *How to Conduct Surveys* by Arlene Fink and Jacqueline Kosecoff, or *Measuring Customer Satisfaction* by Bob E. Hayes (see Appendix B: References and Resources).

The primary focus of process improvement is to uncover drivers of performance, that is, the processes, practices, and other factors that determine performance; it is not just the search for performance measures. Questionnaires should reflect the priority of seeking this cause-and-effect relationship.

Questionnaires that address processes will likely require open-ended questions that ask for Flowcharts. These Flowcharts should at a minimum include sequence of activities, responsibility for activities, and the time for each activity. The next level of detail would include the value created during each activity (i.e., value-added versus non-value-added steps), activity cost, identification of idle time, and possible steps requiring data collection and process improvement.

These guidelines refer to either verbal or written Questionnaires; readers should use their own judgment when applying the guidelines.

General Guidelines

- All Questionnaires should include an introduction that defines the purpose of the Questionnaire and the benefits to the respondent. In many cases, a copy of the study is often sufficient inducement to obtain cooperation. Do not overestimate the motivation of the respondents.

- The sequence of questions asked can have an impact on the validity of the answers. Do not, however, worry about the sequence when the questions are being developed, because the order can be changed before the Questionnaire is finalized. Instead, concentrate on the data requirements of the study.

- Questionnaires should be pretested before being used. Team members should answer the questions for their own organization and modify questions that are difficult to answer. Pretesting should include other internal organizations as well as one or two external sources.

- Instructions should be geared toward the least sophisticated respondent without offending the most sophisticated.

Instructions

- Keep instructions clear and simple. Use boldface, underlines, and/or italics to highlight key points.

- Use brackets or some other convention to separate questions from instructions for interviewers conducting telephone and personal interviews.

- Instructions for questions using a rating scale for answers need to define:

  o   Number of responses expected for each question

  o   Criterion, or standard, for answering the question

  o   Response requested, such as check mark, circle, number, etc.

Organization

- There are three basic parts to a Questionnaire: introduction, body, and conclusion.

- The introduction usually identifies the purpose and scope of the study, provides necessary instructions, and thanks the participants for cooperating. Assurances of confidentially are often included where appropriate.

- The body of the Questionnaire should begin with general, non-threatening questions, followed by more threatening and sensitive ones. Additionally, questions in the body flow from easy, objective, and familiar questions to questions that are more difficult, subjective, and unfamiliar.

Open-ended questions are usually the last ones included in the body. This pattern is usually repeated in each section of the body if it is separated into sections (see note on sections below).

- The conclusion generally includes demographic questions that provide background about the organization.

- Questions throughout the Questionnaire should follow a logical sequence, and transitions between questions and sections should be smooth. Questions should be grouped into sections categorized by topic, issues, or scaling techniques (see Types of Questions below) in order to facilitate responses.

Question Attributes

- All questions on a Questionnaire should be focused on a single issue or thought, the meaning of each question should be clear, and the questions should be as brief as possible. The vocabulary used should avoid abbreviations and industry- or company-specific jargon.

- The most effective sentences are simple sentences (subject and predicate, and sometimes an object). Complex sentences (simple sentence with a dependent clause) are next in terms of effectiveness. These sentence structures are preferable to compound sentences (two simple sentences linked by a conjunction) or compound-complex sentences.

- Questions should not contain words or phrases that will bias the respondent. Wherever possible, middle-of-the-road and neutral wording should replace extreme and sensitive wording. Leading and loaded questions that exhibit a bias should be avoided.

Types of Questions

- There are two basic types of questions: open ended (unstructured) versus forced choice (structured). Most Questionnaires will use both types. Fill-in-the-blank questions are generally classified as unstructured questions.

- Structured questions are generally preferred in many situations because all valid responses are included in the question and results can be compared more easily. Structured questions require more time to develop, however. Both types of questions should be prepared using the question attributes above.

- Structured questions are categorized by the type of response. For process improvement purposes, most questions will be yes–no (one of two or three answers), multiple choice (one or more choices), and scaled response questions, which contain predefined measurement scales.

- Multiple choice questions require that all valid responses be included and that categories of responses be mutually exclusive (i.e., without overlap). In order to ensure that these questions are not too difficult, the number of categories should be limited to ten. Very broad or narrow categories should also be avoided.

- Branching questions are those that instruct the respondent to either go to another question based on his or her answer (conditional) or go to another question and skip sections (unconditional). The sections skipped would normally include questions answered as a result of prior conditional branches. Branching should be limited in order to avoid confusion and possible skipping of important questions.

Unstructured Question: Examples

- **Open-Ended:** These questions are used in the body of a Questionnaire to obtain details about practices, processes, and structural factors that can only be described by the respondent.

  Example: What practices define your activity in this area?

  Example: If yes, briefly describe your process and key measurements.

- **Fill-in-the-Blank:** Fill-in-the-blank questions used in process improvement generally require the respondent to provide specific data. The answer is either an exact or estimated number, percentage, or statement. These questions are often replaced with multiple choice questions to simplify a Questionnaire. If respondents are motivated to answer unstructured questions, then there is no need to spend extra effort in developing structured questions.

  Example: What is your manufacturing lead time? _____ weeks

  Example: What percent of your total engineers are process engineers as opposed to product engineers? _____ % process engineers

Structured Question: Examples

- **Yes–No:** These questions are a simple version of the single-response, multiple choice question. Valid answers are either "yes," "no," or, in some cases, "don't know." These simple questions are often used in the body of a Questionnaire to test for the presence of practices, processes, and structural factors. They will often include conditional branching if there is a positive response, which will then request, for example, a description of a process (see second open-ended example above), a list of practices, a list of other factors, or specific data and information (fill-in-the-blank).

  Example: Do you have a formal system or process in place for measuring customer satisfaction? Yes ___ No ___

- **Multiple Choice:** These relatively simple questions can have either a single or multiple response from among a limited number of discrete categories (maximum of ten). Multiple choice questions are commonly used throughout all sections of a Questionnaire. For process improvement purposes, the discrete categories are generally numerical intervals, ordinal ranking, or descriptive categories. Intervals are often used if precise measures are not needed, the data are sensitive, or precise data would be difficult to obtain.

  Example: What percent of your suppliers use Statistical Process Control?
  10% to 15% ___   15% to 20% ___   etc.

  Example: How many layers of management are there in the R&D organization? 2 ___ 3 ___ 4 ___ 5 ___ 6 ___ 7 ___ greater than 7 ___

- **Scaled Response:** Scaled response questions are relatively easy to answer and are often used in cases where many similar questions can be answered using the same scale.

  o **Frequency:** When exact percentages are not needed or are difficult to calculate, scaled frequency response questions often suffice. The frequency categories can be increased to provide further definition. The use of an even number of categories forces answers out of the neutral position ("fence riding"). The following example assumes that this is not a problem.

    Example: How often do the following organizations become involved in product design? (Check the appropriate column for each organization.)

	Never	Sometimes	Always
Mfg. Operations	❒	❒	❒
Purchasing	❒	❒	❒
Quality Control	❒	❒	❒
Test Engineering	❒	❒	❒
Suppliers	❒	❒	❒
Customers	❒	❒	❒
Marketing	❒	❒	❒

  o **Forced Ranking:** When the relative standing of categories is required and the interval between ranked categories and the absolute value is not important, then forced ranking is used. The number of categories is usually limited to ten items in order to avoid making the question too difficult. A single judgment criterion must be clearly stated to avoid confusion.

    Example: What measures of performance, in order of importance, do your customers use to measure their satisfaction with your products and services? (Rank in order of importance, with 1 as most important and 5 as least important.)

Product reliability	____
Cost of ownership	____
Product features	____
Repair cost	____
Order lead time	____
Service response time	____

o **Horizontal Numeric Scale:** When items are to be judged on a single dimension with a scale of equal intervals, then a simple horizontal numeric scale can be used. The use of an even number of categories forces answers out of the neutral position ("fence riding"). The following example assumes that this is not a problem.

Example: Rate the following service practices according to how important the service is to your overall satisfaction and how effectively the service is currently being performed. (Check the appropriate column for importance and the appropriate column for effectiveness for each service.)

	Extremely Important				Very Unimportant	Effective				Ineffective
	5	4	3	2	1	5	4	3	2	1
Scheduled times	☐	☐	☐	☐	☐	☐	☐	☐	☐	☐
Easy credit	☐	☐	☐	☐	☐	☐	☐	☐	☐	☐
800 number	☐	☐	☐	☐	☐	☐	☐	☐	☐	☐

o **Comparative Scales:** Comparative scales are used to judge a standard entity with other items in order to distinguish variances above or below the standard. Respondents need to be familiar with the standard in order for results to be valid. This method produces both interval data and relative ranking.

Example: Compared to previous product design practices, rate the following new practices. (Check the appropriate column for each practice.)

	Very Superior	About the Same			Very Inferior
	5	4	3	2	1
Expanded design teams	☐	☐	☐	☐	☐
Use of CAD equipment	☐	☐	☐	☐	☐

CHAPTER FIFTEEN

VIDEO ONE
CASE STUDY

The examples in this book are based on the Video One Case Study presented in this chapter. Although each tool and technique is described in detail in earlier chapters, the case study provides a context for applying them that often aids learning. Therefore, it is recommended that readers complete this chapter before exploring the tools in Chapters Five through Twelve.

The Storyboard technique presented in Chapter Eleven follows the Video One Case Study, but the tools and techniques vary from the ones presented here. The primary reasons are to:

- Illustrate the use of different tools and techniques to achieve similar results. There is no cookbook approach to quality improvement.

- Demonstrate that Storyboards highlight key findings and recommendations of a single project but do not provide the background detail of a case study.

INDUSTRY BACKGROUND

The home video business continues to show strong growth, primarily at the expense of network and pay TV channels. Revenue growth exceeds 20% per year in most markets, and current industry forecasts indicate that this growth will continue for the next five years. Retail revenue projections for 1995 exceed $20 billion, which is three times projected domestic movie box office revenues. Video rentals are 65% of total retail revenues. Declining prices on videotapes for sale, stable prices on rental tapes, and rapid growth in titles released drive this growth.

The majority of video consumers are "baby-boomers," characterized by more leisure time, higher disposable income, and fewer children. A recent survey of video retailing found that 75% of customers of video stores were in the 25- to 45-year age group. The average customer's annual income was $27,500. Video consumers are 55% male and 45% female.

The industry is reaching a saturation point as the number of outlets exceeds 20,000 nationwide. Current predictions indicate that a quarter of these outlets will go out of business or be acquired as the industry consolidates in the next five years. Larger outlets will likely dominate local markets. More than half of the current stores have a competitor within one-quarter mile of their store.

VIDEO ONE

The Vinyl Corporation, a large multinational conglomerate, acquired your company, Video One, seven months ago. Video One has ten competitors in Hicksville, a small but growing suburb of a large metropolitan area. Your store is 1700 square foot in a building that has ample parking. Expansion at your current location is possible.

In recent years, your annual revenue growth exceeded 20%, but current projections are 5% per year. Total revenue estimates for this year are $330,000. Profits have also declined from a return on sales of 18% two years ago to a forecasted 12%. Your business consists of the rental and sale of prerecorded videotapes. You do not currently sell blank tapes, nor do you rent or sell equipment.

Since Blockbuster opened four years ago, you have noticed that your average customer is getting older, and you estimate that the average age is now 45 years old. You have developed a loyal customer base, but have had little success attracting new customers. Your experience indicates that pricing, title selection, and convenience are important to your customers.

You sell annual memberships for $10, which entitles members to a $0.50 discount off your regular rental price of $2.50 per day. Through the buying power of the Vinyl Corporation, however, your cost per title is as low as your largest competitor, Blockbuster.

Vinyl Corporation's acquisition strategy assumed that Video One could achieve previous profitability levels. Mr. Fetish, President of Vinyl Corporation, is conducting his first Operations Review with the staff of Video One. His primary objectives for this meeting are to get a status on the current performance and to identify opportunities for improving results.

As a strong advocate of Total Quality Management, the Vinyl Corporation provided all Video One employees with 40 hours of quality management training. This training included quality philosophy and various tools and techniques necessary for the employees to work as a team in solving problems and pursuing improvement opportunities.

VIDEO ONE'S FIRST OPERATIONS REVIEW

The staff of Video One present at the meeting include:

Geraldo, the new Store Manager

Ernestine, the Cashier

Willard, the Inventory Clerk

Ivan, the Accountant

Buffy, the Marketing/Purchasing Manager

Ernestine started working for Video One when it opened in 1985. Willard joined Video One in 1987, and all other employees were hired since the Vinyl Corporation acquisition.

Mr. Fetish requested that Ivan present several performance measures for discussion at the meeting. Ivan prepared the following data:

Average Inventory-For Sale	300
Average Inventory-For Rent	2,500
Rental Titles	1,000
Average Sale Price	$32
Average Rental Price	$2.10
Annual Sales Revenue (12-month projection)	$57,600
Annual Rental Revenue (12-month projection)	$273,000
Full-Time Employees	8
Check-Out Time (minutes)	6 to 10

Before the acquisition, Mr. Fetish analyzed Video One and its competitors in Hicksville. Mr. Fetish presents a portion of this data to help the team understand the predicament it faces.

TABLE 15.1   Video One Data Table

	20/20	Blockbuster	Popingo	Wherehouse	Video One
Sale Inventory Turns/Mo..	1.0	2.0	0.5	0.7	0.5
Rental Titles	600	2,000	500	1,000	1,000
Average Copies/Title	3.3	4.0	2.0	3.0	2.5
Rentals/Week/Copy	1.5	2	1.2	1.5	1
Average Price					
Sale	$30	$24	$33	$28	$32
Rental	$2.50	$2.00	$3.00	$2.00	$1.80
Annual Units (000)					
Sale	4.8	12.0	1.2	3.2	1.8
Rental	156.0	832.0	62.4	234.0	130.0
Annual Revenue ($000)					
Sale	$144	$288	$40	$90	$58
Rental	390	1,664	187	468	234
Total	$534	$1,952	$227	$558	$292
Market Share ($)	11%	41%	5%	12%	6%
Revenue Growth/Year	10%	20%	-5%	10%	5%
Return on Sales	15%	25%	3%	21%	12%
Full Time Employees	10	25	6	7	8
Sales ($000)/Employee	$53	$78	$38	$80	$36
Average Staffing	4.8	11.0	2.9	3.3	4.6
Store Size (Sq. Ft.)	1,200	4,000	750	1,700	1,700
Sales ($)/Sq. Ft.	$445	$488	$302	$328	$172
Rental Copies/Sq. Ft.	1.7	2.0	1.3	1.8	1.5
Computerized Check Out	Yes	Yes	No	Yes	NO
Drop Box (After Hours)	Yes	Yes	Yes	NO	NO
Hours of Operation	10AM-10PM	10AM-11PM	9AM-9PM	10AM-10PM	10AM-8PM
Location	Strip Mall	Near High School	Strip Mall	Mall	Own Bldg.
Membership	Yes	Yes	Yes, $30/Yr.	No	Yes,$10/Yr.
Member Discount	None	None	20%	None	$0.50/Rental
Parking	Adequate	Ample	Difficult	Adequate	Ample
Parking Spaces	10	50	3	15	25

SELECTING AN IMPROVEMENT OPPORTUNITY

After reviewing the status of the business, the staff begins to explore improvement opportunities. Mr. Fetish is aghast at the long check-out time, and this subject consumes much of the discussion at the meeting. Mr. Fetish facilitates the discussion and suggests that the team brainstorm other problem areas and opportunities for improvement. The team develops the following list of items, which have been combined and prioritized using List Reduction. (As a facilitator,

Mr. Fetish helps the team with the meeting process, but he does not make suggestions regarding the content of the discussion.)

Problems/Improvement Opportunities

1		[Low return on sales]
2	A	Long check-out time
	A	Check-out process not efficient
3	B	Not enough titles
4		Store layout poor
5		[Insufficient display space]
	B	Wrong movies in stock
	B	[Too few science fiction movies]
	B	Too few copies of new releases
6		[Declining profits]
	B	Poor selection
		~~Too long to check out~~
7		[No drop box for after hours]
8		No computer controls

FIGURE 15.1   Brainstorming: Problems/Improvement Opportunities

After Weighted Voting is applied to the remaining alternatives, the team reduces the list to four items:

Check-out-related items ("A category")

Title selection items ("B category")

Store layout (#4)

No computer (#8)

Mr. Fetish suggests that the team try to relate each problem/improvement opportunity to a key process that influences performance in the areas identified. After much discussion, the team reaches consensus on the following points:

- Store location is a "structural" factor that cannot be changed and must be accepted as is.

- Use of manual methods (i.e., a practice) rather than mechanization using a computer will be addressed within specific processes.

- The team identifies two key processes involved: Check Out and Title Selection. On further discussion, the team agrees to consider two other related processes: Advertising and Survey Design and Analysis.

The team then uses Paired Comparisons to reach consensus on what members feel is the most important area needing improvement: the Check-Out Process.

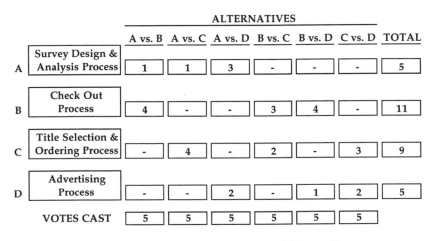

		ALTERNATIVES						
		A vs. B	A vs. C	A vs. D	B vs. C	B vs. D	C vs. D	TOTAL
A	Survey Design & Analysis Process	1	1	3	-	-	-	5
B	Check Out Process	4	-	-	3	4	-	11
C	Title Selection & Ordering Process	-	4	-	2	-	3	9
D	Advertising Process	-	-	2	-	1	2	5
	VOTES CAST	5	5	5	5	5	5	

FIGURE 15.2 Paired Comparisons: Comparison of Alternatives

Check-out time is one of several revelations for Mr. Fetish during the meeting, however. In addition, he is somewhat concerned about the method that Video One uses to conduct its market research, how it surveys its customers, and the method used in selecting and ordering tapes for the rental inventory. Buffy, recently hired to manage both marketing and purchasing functions at Video One, performs most of these duties. Previously, Ernestine performed many of these duties.

After discussing various proposals, the team agrees to the following problem/opportunity statement:

As Is: Check-out time of 6 to 10 minutes is too long

Desired State: Reduce check-out time by 50% in six months

At the end of the meeting, the team agrees to perform various tasks for the next monthly Operations Review. The entire team agrees to prepare a Deployment Flowchart to document the Check-Out Process. In addition, Buffy agrees to prepare a short Questionnaire to solicit customer feedback on suggestions and complaints. The form will be available at several locations in the store and will be collected using a suggestion box located next to the cashier. Buffy will tabulate the results for the next meeting.

Finally, Geraldo will start random sampling of check-out times to establish a baseline of performance. He plans to select random times during the day (using a random number table) to measure actual check-out times.

ANALYZING CURRENT ACTIVITY

During the second Operations Review, the team displays a Deployment Flowchart (Figure 15.3) of the Check-Out Process. The team prepared this chart in conjunction with a Process Analysis Worksheet. The worksheet begins the process definition activities and includes customer–supplier data, input data, a Top-Down Flowchart, and a Table of Performance Indicators and Drivers.

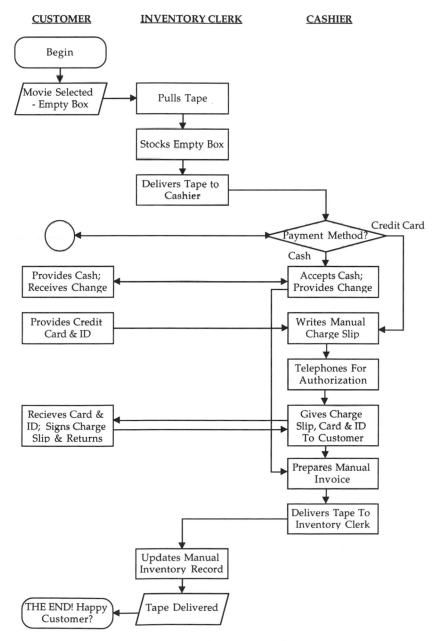

FIGURE 15.3 Deployment Flowchart: Video One Check-Out Process

After some discussion, the team agrees that the Deployment Flowchart correctly represents the steps in the process, with the exception of the problem that Willard brought up. It seems that some customers, particularly some of Ernestine's friends, do not rewind the tapes and Willard often must rewind them while customers wait. Ernestine interjects that most of her friends are senior citizens and sometimes forget to rewind the tapes, but they are good customers in spite of their forgetfulness. Ernestine explains to Mr. Fetish that the long check-out time gave her an opportunity to gossip with her old friends.

Buffy presents the following summary of customer complaints.

	Mon.	Tue.	Wed.	Thur.	Fri.	Sat.	Sun.	Total
Check Out Time	✓✓	✓✓✓	✓✓✓	✓✓✓✓✓ ✓✓✓	✓✓✓✓✓ ✓✓✓✓✓ ✓✓	✓✓✓✓✓ ✓✓✓✓✓ ✓✓✓✓	✓✓✓	45
No New Releases		✓✓			✓✓✓✓✓ ✓✓	✓✓✓✓✓ ✓✓✓	✓✓✓	20
Poor Title Selection		✓✓✓			✓✓✓✓	✓✓✓✓✓ ✓✓✓		15
Can't Find Movies				✓✓	✓✓✓	✓✓	✓	8
Other		✓	✓	✓✓			✓✓	6
	2	9	4	12	26	32	9	94

FIGURE 15.4   Check Sheet: Customer Complaints

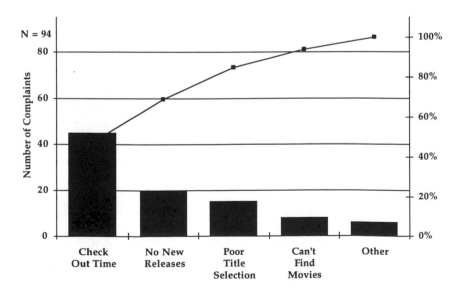

FIGURE 15.5   Pareto Chart: Customer Complaints

The team prepared a Pareto Chart during the meeting to highlight the importance of check-out time and to validate their improvement opportunity.

Geraldo shows the following Run Chart to the team. He calculated the average check-out time to be 8 minutes. During the past month, Geraldo visited competing stores and observed their check-out activities. He shares his observations and his estimates of competitor check-out times: Popingo, 10 minutes; 20/20 and Wherehouse, 5 minutes; and Blockbuster, 3 minutes.

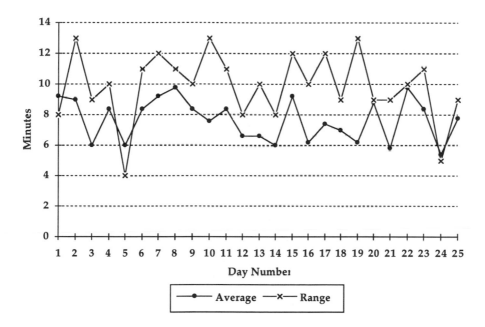

FIGURE 15.6   Run Chart: Check-Out Time

For the next meeting, Buffy agrees to report on a Focus Group meeting of customers organized to better understand customer needs and expectations and potential reasons for customer dissatisfaction with Video One. The group will include both Video One and Blockbuster customers; participants will be offered $15 worth of free video rentals at Video One as an incentive to participate. Buffy will seek customer comments on the differences in Check-Out Process used by Blockbuster versus Video One's process. Geraldo and Willard will begin recording all check-out times and will report their findings at the next meeting. Ivan agrees to develop a Work Flow Diagram of a typical rental transaction. He feels that the process is not convenient for customers.

The team decides to revise the problem/opportunity statement as follows:

As Is: Check-out time averaging 8 minutes is the primary reason for customer complaints

Desired State: Reduce check-out time by 50% in six months and reach benchmark standard in one year

As the meeting concludes, the team briefly discusses the changes in the benefits plan that Mr. Fetish outlined at the beginning of the meeting, including his desire to increase wages and provide training in customer service and cross-training of all store functions. Both Ernestine and Willard oppose cross-training, but agree with all the other changes proposed by Mr. Fetish.

IDENTIFY ROOT CAUSES

During the third Operations Review, Buffy discusses the Interrelationship Digraph (Figure 15.7) that her customer Focus Group developed. This chart shows that check-out time is not Video One's only problem. It also highlights the role of other factors as possible root causes of customer dissatisfaction with Video One. These include manual invoicing, manual inventory control, lack of training, and tapes not rewound.

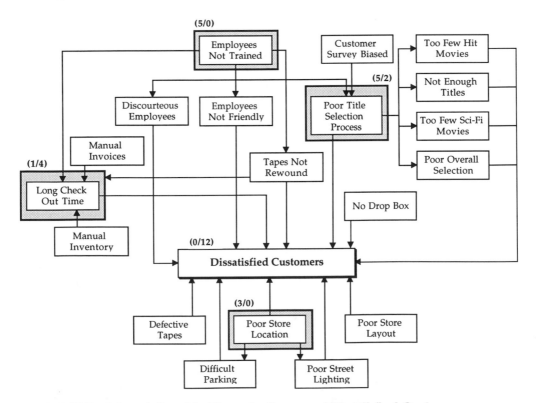

FIGURE 15.7 Interrelationship Digraph: Causes of Dissatisfied Customers

Willard and Geraldo present Control Charts (Figures 15.8 and 15.9) to further document the existing baseline performance. The average and range Control Charts show that the current process average of 7.7 minutes is lower than the first estimate, but is still significantly higher than Blockbuster's average. The wide fluctuation in check-out times in the range Control Chart shows that the existing process varies significantly. This variation likely depends on the number of tapes rented in each transaction.

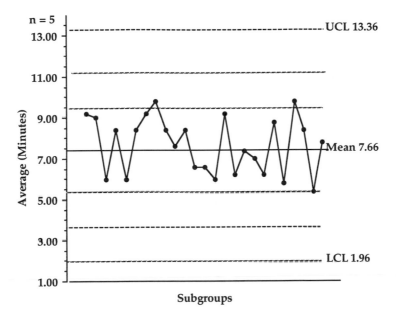

FIGURE 15.8 $\overline{X}$ **Control Chart: Average Check-Out Time**

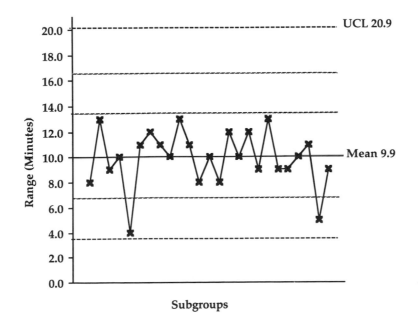

FIGURE 15.9   **R Control Chart: Check-Out Time Range**

Finally, Ivan presents the Work Flow Diagram, which illustrates another problem with the current Check-Out Process: customers must walk back to the inventory clerk's window to pick up tapes after paying the cashier.

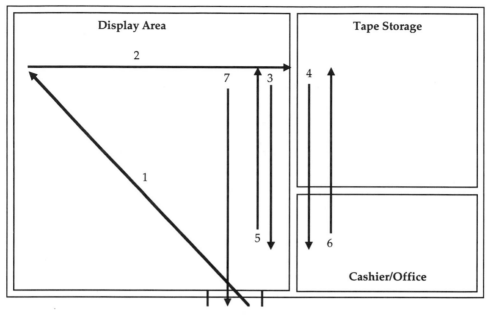

FIGURE 15.10 Work Flow Diagram: Video One Check Out

Ivan explains the Work Flow Diagram as follows:

Step	Activity
1	Customer enters and makes selection
2	Customer gives tape to the Inventory Clerk
3	Customer goes to Cashier to wait for tape and pay Cashier
4	Inventory Clerk pulls tape and delivers to Cashier (some tapes must be rewound during this step)
5	Customer walks back to Inventory Clerk's window
6	Cashier returns tape to Inventory Clerk to update inventory records
7	Customer receives tape and leaves store

For the next meeting, Geraldo and Willard agree to continue collecting check-out time data, but will now address actual and theoretical times for completing each step in the process.

SELECT AND PLAN SOLUTION

The team's fourth meeting begins with a lively discussion of the revised Deployment Flowchart (Figure 15.11), which includes Geraldo and Willard's estimate of theoretical and actual times by process activity. The team brainstorms a list of potential improvements in the process. After List Reduction, the team inserts the best ideas into the "Quality Issues" column of the Flowchart, as shown.

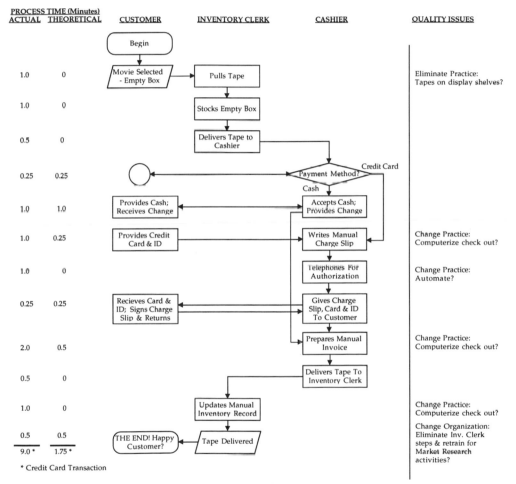

FIGURE 15.11   Deployment Flowchart: Video One Check-Out Process

The team's productivity and energy in tackling the check-out time project impresses Mr. Fetish. He immediately approves the team's recommended solutions proposed in the "Quality Issues" column of the Flowchart.

Before the meeting ends, Geraldo agrees to work with the team to prepare a written implementation plan for Mr. Fetish's approval. This detailed proposal will include a description of major changes in the Check-Out Process; computer software evaluation; responsibility for implementing changes; estimate of the capital costs for remodeling the store, computer equipment, and software; and a milestone completion chart.

DURING IMPLEMENTATION

Mr. Fetish approved the detailed implementation plan upon receipt one month ago. We now find the team reviewing its progress on the Check-Out Project during the monthly Operations Review. The team uses a Gantt Chart (Figure 15.12) to highlight major milestone activities.

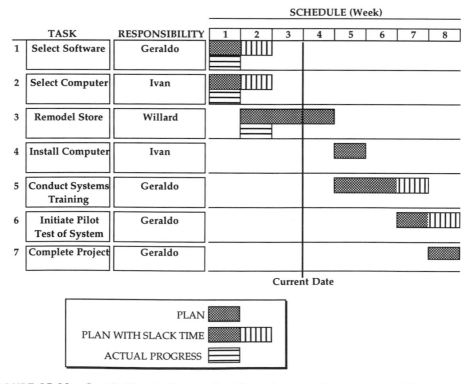

FIGURE 15.12   Gantt Chart: Check-Out Time Process Improvement Project

Mr. Fetish is pleased with the team's progress and is assured by the team members that they will continue tracking check-out time and will report improvements in subsequent Operations Reviews. Mr. Fetish suggests a Grand Reopening celebration to recognize the team's achievements.

Before the meeting concludes, Buffy and Willard propose that the team start a new project immediately after the reopening: Title Selection Process Improvement Project.

APPENDIX A

GLOSSARY

Activity Network Diagram: A management and planning tool used to develop a time–sequential plan for implementing projects or improvements.

Affinity Diagram: A management and planning tool used to translate large numbers of complex, apparently unrelated ideas, issues, or opinions into natural and meaningful groupings of data.

Agenda: A key meeting management tool to ensure that meetings are effectively planned and managed.

"As Is" Statement: *See* Problem Statement.

Attribute Data: Data obtained by counting (discrete interval data), expressed in whole numbers without fractions or decimals. Used in Control Charts that show either non-conforming items (defects) or the occurrence of non-conformance (defective). Also referred to as discrete data.

Balance Sheet: A tool used by teams to identify the pros and cons of alternatives in an attempt to reach consensus.

Benchmarking: An externally focused quality improvement technique that seeks the "best" practices of others (other divisions within the organization, competitors, or organizations that perform functionally equivalent activities). Once superior performance is identified and drivers of that performance are understood, the "best" practices can be applied by all organizations participating. Benchmarking provides something for everyone, because it is unlikely that one organization does everything well or cannot learn from others.

Brainstorming: An idea generation technique that stimulates group creativity and generates many ideas in a short time period. Brainwriting and the Crawford Slip Method are two alternate Brainstorming techniques that rely on written communication rather than the verbal communication in Brainstorming.

c Chart: Control Chart for plotting data based on the total count of non-conformities (defects) with a sample size that remains constant.

Cause: A proven reason for some measured or observed effect, often a failure or defect. *See also* Root (or Primary) Cause.

Cause-and-Effect Diagram: A graphic technique used to identify and relate causes and subcauses with a particular effect or result. Also referred to as an Ishikawa diagram or fishbone diagram.

Charts (Bar, Pie, Run, and Spider):   Graphic tools used to visually display data that aid understanding and decision making by providing a clear summary of the data, showing relative importance of the data elements, and illustrating patterns of relationships often lost in tabular presentations of data.

Check Sheet: Data recording table used to measure the occurrence of specific events during a data collection period.

Common Cause: A cause of variation in a process that is random and uncontrollable because it is inherent in the process.

Consensus: A condition reached when a group has considered all aspects of an issue, when all members of the group have voiced their opinion, and where each member accepts, supports, and does not oppose a decision to be made.

Continuous Improvement: Operational philosophy that makes the best use of resources in order to increase product or service quality and result in more effective satisfaction of customers.

Control Charts: Charts that show the sequential performance of a process relative to its expected level of performance. Used to determine when a process is operating in or out of statistical control as defined by control limits.

Control Limits: Statistically derived limits for a process that indicate the spread of variation due to natural, random variation in the process. Control limits are based on averages of a series of sample measurements.

Crawford Slip Method: *See* Brainstorming.

Criteria Rating Form: Tool used to rate alternatives and reach consensus through the application of various weighted criteria. *See also* Prioritization Matrix.

Critical Success Factors: Factors based on customer knowledge that indicate what the organization must focus on in order to satisfy customer needs and expectations.

Cross-Functional: Involvement of multiple functions within an organization. Usually used in describing quality teams that address processes that "cross" functions.

Customer: Anyone who consumes or uses a product or service, whether internal or external to the organization or the individual providing the product or service.

Customer (Internal): Customer inside the organization who receives outputs from other individuals or departments within the organization. Owners and shareholders are often assumed to be internal customers.

Customer Needs and Expectations: Customer wants, desires, wishes, and demands for products and services translated into measurable indicators of cost, quality (various dimensions), and delivery (time and quantity). Often referred to as customer requirements.

Defect: A non-conforming attribute.

Defective: A unit with non-conforming attributes, or defects, that make the item not acceptable.

Deployment Flowchart: A flowcharting method that shows not only the sequence of activities and decisions, but also the responsibility and/or location for performing each step in the process.

"Desired State": *See* Problem Statement.

Diagnosis: The process of studying symptoms, collecting data, and conducting experiments to test theories and establish relationships between causes and effects.

Effect: An observable condition or measurable evidence that indicates a problem, often a failure or defect.

Effectiveness: A condition where the quality and quantity of an output from a process satisfies a customer. Effectiveness results from "doing the right things" (judgment) and "doing things right" (execution). The primary measure of effectiveness is external customer satisfaction, a desired outcome. Internal customer (and stakeholder) satisfaction is often measured as business or organization effectiveness. *See also* Outcome.

Efficiency: A measure of performance where the output of a process is measured in terms of the inputs to that process (i.e., outputs divided by inputs). Processes that produce outputs whose value exceeds the costs of inputs by the greatest amount are the most efficient.

Facilitator: A person who manages a group or team process while allowing the group or team leader to focus on the content of activities performed.

Fishbone Diagram: Another name for Cause-and-Effect Diagrams. The completed diagram resembles a fish skeleton.

Five Whys: A technique used in cause-and-effect analysis where the search for root causes begins with the question, "Why does this 'effect' occur?" The question is then asked four more times. The technique can be use with Cause-and-Effect Diagrams or with the more structured Tree Diagram approach.

Flowcharts: Graphic representations that symbolically show the sequential activities in a process that produces an output. Various forms exist, ranging from macro-level charts (e.g., Top-Down Flowcharts) to micro-level charts (e.g., Deployment Flowcharts and Work Flow Diagrams), which can also show the type of activity, inputs, outputs, responsibility, spatial relationships, time, costs, and value.

Focus Group: An group interviewing technique used to gather data from customers about their needs and expectations, as well as their suggestions for improving products and services.

Force Field Analysis: A list of helping factors that must be capitalized upon or hindering factors that must be overcome before problems can be solved or opportunities captured. This analysis helps to focus on the most important factors that assist or resist change.

Gantt Chart: A planning tool used to document the schedule, activities, and responsibilities necessary to complete a project or implement a proposed solution.

Goal: A statement that describes a desired future condition or change.

Goal Alignment: A desirable situation in an organization where the goals and objectives of work units are consistent and supportive of the goals and objectives of the entire organization.

Histogram: A Bar Chart that displays the frequency distribution or variation in one characteristic of data.

Improvement Opportunity: An area, topic, or process identified by management, Steering Committees, customers, or teams as needing improvement.

In Control: A process operating with natural, random variation inside the control limits. *See also* Control Limits.

Input: Resource consumed, utilized, or added in a process that produces a product or delivers a service.

Interrelationship Digraph: A graphic technique for determining the cause-and-effect relationship between a given issue or problem and the factors that affect it.

Ishikawa Diagram: Another name for Cause-and-Effect Diagram, named for Japanese engineer Dr. Kaoru Ishikawa, who developed it.

Is/Is Not: A method used to help teams separate or stratify information, data, and issues into like categories and define the categories for further analysis.

Just-in-Time: A philosophy that calls for goods and services to be produced and delivered just prior to their need. The goal is to eliminate waiting time and inventory queues in a process.

List Reduction: A technique for reducing the number of ideas generated into a meaningful and manageable list. The method groups like ideas, eliminates duplicates, and filters the list using simple majority voting to highlight important items.

Management Tools: The "seven management tools" include Activity Network Diagram, Affinity Diagram, Interrelationship Digraph, Matrix Diagram, Prioritization Matrix, Process Decision Program Chart, and Tree Diagram.

Matrix Diagram: A graphic technique used to analyze the relationship between two or more related groups of characteristics, ideas, or issues. The symbols denote the strength of relationships and provide an easily recognizable visual pattern to uncover relationships.

Mean: The average of a data set, that is, the sum of all values divided by the number of data values in the set.

Median: The value half way between the highest and lowest value when all values of a data set are listed in order (either in ascending or descending order).

Meeting Evaluation: A technique used by teams to evaluate the effectiveness of their meetings. *See also* Plus/Delta (+/Δ) Evaluation.

Mission: The single overriding goal statement of an organization that defines its purpose, its philosophy, and its reason for existing. This philosophy explains how the organization values and utilizes its resources.

Mode: The value in a data set that occurs most frequently.

Non-Value Activities: Any activity performed in producing a product or delivering a service that does not add value, where value is defined as something for which a customer would be willing to pay.

Normal Distribution: A symmetrical bell-shaped distribution for variable data in which the mean, median, and mode are all the same. This distribution type is the basis for all variable Control Charts.

np Chart: Control Chart used to plot the number of defective items in a sample where the sample size remains constant.

Objective: A more specific statement of the desired future condition or change than a goal. It includes measurable results to be accomplished, often within specified time limits.

Outcome: A measurable result or accomplishment that indicates whether both the quantitative and qualitative characteristics of a product or service satisfy expectations. Desirable outcomes occur when processes are effective. In a more general definition, an outcome defines a level of success in achieving specific goals and objectives. *See also* Effectiveness; Performance Indicators.

Out of Control: A process with variation outside the control limits caused by special causes. *See also* Control Limits; Special Cause.

Output: A product produced or a service delivered by a process. It is often measured in terms of the quantity and quality of product or service delivered that meets a certain requirement, such as cost, quality (dimensions), and delivery (time and quantity) requirements. *See also* Outcome, which measures the degree to which a product or service satisfies expectations.

p Chart: Control Chart used to evaluate performance based on the fraction or percentage defective in a sample whose size is either constant or varying.

Paired Comparisons: A simple voting technique to help teams identify their preferences and rate various alternatives while they attempt to reach consensus. It consists of all team members voting for one alternative in each pair of alternatives.

Pareto Chart: A vertical Bar Chart that displays the relative importance of categories of problems or conditions. The items are arranged in descending order, from left to right, and help to separate the vital few important items from the less important, trivial ones.

Performance Drivers: The processes, practices, and structural elements that influence performance and are the means for achieving measurable objectives. Drivers define how the organization performs at both the organizational and operational level. Changes in performance drivers must occur before changes in outcomes can be measured.

Performance Indicators: Measures used to indicate levels of performance directly related to customer requirements and supplier specifications. Indicators are both quantitative and qualitative and are expressed in terms of the dimensions of quality, cost and delivery. They are often categorized as either outcomes (defined by customer needs and expectations) or process variables and attributes (defined by supplier specifications).

Plan-Do-Study-Act Cycle: The sequence of activities used in problem solving or process improvement where planning is followed by implementation (doing, often in pilot activities), analysis of results (studying, monitoring, or evaluating), and standardization (action), which is then repeated for continuous improvement. Also referred to as the PDCA cycle ("C" for Check), PDSA cycle, Shewhart cycle (named after Statistical Process Control pioneer Dr. Walter A. Shewhart), or Deming Wheel (named after Dr. W. Edwards Deming, a Shewhart disciple who led the third wave of the Japanese industrial revolution).

Plus/Delta (+/Δ) Evaluation: A technique used by teams to evaluate the effectiveness of their meetings. This critiquing tool identifies what is working well (primarily the meeting process) and what needs improvement. *See also* Meeting Evaluation.

Practices: The policies, procedures, and methods used to perform activities. They are integral to a process and often define how work is performed. Documentation of practices and the development of Flowcharts to understand the sequence of, timing, and responsibility for activities are the two most important activities in understanding processes.

Prioritization Matrix: Management tool used to rank ideas and to select tasks or projects based on weighted criteria as determined by group consensus. The complexity and levels of detail handled by this tool limit its use. Criteria Rating is a simplified version of this tool.

Problem Statement: A statement that describes the current condition, or problem, and the desired condition in specific measurable terms. This two-part statement defines the "as is" condition and the "desired state" in unbiased specific terms, where potential causes or solutions are not identified.

Process: A series of activities that transforms and adds value to various resources and inputs in order to create outputs used by either internal or external customers.

Process Capability: A technique used to determine whether a process, given its natural variation, is capable of meeting customer specifications for a product or service.

Process Decision Program Chart: A management and planning tool used to create a detailed implementation plan that includes all conceivable unfavorable events and possible problems that may occur.

Process Owner: A manager or supervisor assigned responsibility for a process because of his or her influence over the process, either as a major contributor, decision maker, or someone whose performance is measured in terms of the success of the process.

Process Stakeholder: All members of the work unit who are involved with a process.

Process Variables and Attributes: A series of quantitative and qualitative performance indicators that describe a process that produces products or services. They include inputs, outputs, activity levels, utilization levels, and efficiency measures, expressed in terms of quality, cost, and delivery measures. They are often defined by supplier specifications. *See also* Performance Indicators; Supplier Specifications.

Productivity: A measure that combines the efficiency of a process (i.e., output per unit of input) and the utilization of the process (i.e., actual activity relative to theoretical activity). A process that produces efficiently with a high utilization is productive. *See also* Efficiency; Utilization.

Quality: The totality of features and characteristics of a product or service that bear on its ability to satisfy customer needs and expectations.

Quality Council: A senior management team responsible for establishing the quality improvement system policies and plans. These policies and plans are then deployed by a steering committee representing functional departments that guide the implementation. *See also* Steering Committee.

Quality Improvement: A systematic method for improving processes, practices, and structural factors in order to better meet customer needs and expectations.

Quality Tools: The "seven quality tools" include Check Sheets, Cause-and-Effect Diagrams, Charts, Histograms, Pareto Charts, Scatter Diagrams, and

Control Charts. Cause-and-Effect Diagrams are more analytical than statistical, but their usage is closely linked to these statistical tools.

R Chart: A Control Chart that plots the range of sample data values as a function of time, sequence, or lot number.

Random Cause: A natural variation that is inherent in a process. *See also* Common Cause.

Random Sample: A sample chosen from a larger population by a method that gives each unit an equal chance of being selected.

Range: A measure of variability in a data set calculated by subtracting the lowest value from the highest value.

Rework: Activity required to correct for defects produced by a process.

Road Map: Any written documentation used to plan or report progress of a project.

Root (or Primary) Cause: The basic reason, or causal factor, that results in some measured or observed effect, often a failure or defect. The root cause in many cases consists of several smaller causes.

Run Chart: A graphic plot of a measurable characteristic of a process versus time.

Sample Size: Number of items selected in a sample. The size of the sample is determined by the purpose of the study and the acceptable levels of accuracy required. For most Statistical Process Control, the sample size remains constant each time samples are selected.

Sampling: A technique for obtaining information about a large group (the population) using a smaller representative group (the sample). The population may be finite (that is, from a specific lot) or infinite (from a continuous process). Sampling is categorized by how the sample is drawn, such as random, stratified, clustered, or two-stage.

Scatter Diagram: A chart in which one variable is plotted against another to determine if there is a correlation between the two variables. Scatter Diagrams can be used to verify cause-and-effect relationships, but a strong correlation does not necessarily indicate such a relationship because a third factor may be the causal factor.

Scribe: A person who writes inputs from a team on flip charts or note cards during meetings.

Special Cause: A cause of variation in a process that is not random and that is controllable. It is caused by controllable forces not defined by or inherent in the process.

Specification Limits: Limits established for a process either defined by the customer or determined by the supplier to satisfy customer needs and expectations. These limits are applied to individual occurrences and are not necessarily consistent with the control limits of the process producing the product or service. *See also* Process Capability.

Stable Process: A process in statistical control. This means that all points on a Control Chart fall within the 3σ limits and that no statistically improbable patterns occur.

Standard Deviation: A mathematical term to express the variability in a data set or process. It is represented by the lowercase Greek letter sigma (σ). In a normal distribution, 68.26% of the data will be within 1σ from the mean, 95.45% within 2σ, and 99.73% within 3σ. Mathematically, a standard deviation is equal to the square root of the average squared differences between individual data values and the data set average. (Note that σ, which is the population standard deviation, has been substituted throughout this book for the sample standard deviation, which is denoted by the letter "s." This is consistent with current practices.)

Statistical Process Control (SPC): The application of statistical methods to monitor variation in a process. The shift from Statistical Quality Control to SPC parallels the shift from inspection by quality control to prevention by everyone involved with the process.

Statistics: The study of numbers or summary measures calculated from a set of data.

Steering Committee: A middle management/supervisor team responsible for deploying the quality improvement system policies and plans throughout the organization. The committee represents all functional departments and guides the quality implementation defined by the Quality Council. *See also* Quality Council.

Step Zero: The development of competitive and strategic assessments where the organization gains knowledge about its customers and competitors, develops strategies to achieve competitive advantages, and begins to understand itself in terms of its strengths and weaknesses. In addition, organizations often begin to document processes (with Flowcharts) and practices (with written operating procedures) during step zero.

Storyboards: A planning tool for providing an organized visual summary that is useful in planning quality improvement projects as well as documenting and communicating results. Each visual frame shows the chronological and logical flow of the activity being planned or described.

Subject Matter Expert (SME): A person with thorough knowledge and experience in specific functional activities within processes.

Subprocess: A group of activities that together accomplish a significant portion of an overall process.

Supplier: Anyone from whom an organization or individual receives a product or service.

Supplier Specifications: The translation of customer needs and expectations, or requirements, into measurable indicators of quality, cost, and delivery (time and quantity) understood by the supplier.

Surveys: A means (either written using Questionnaires or verbally during interviews) of gathering data on the opinions, feelings, impressions, or satisfaction of a group. Also used to gather data for quality improvement of products, services, functions, processes, and practices.

Team Leader: A person who leads a team through the quality improvement process.

Team Member: A person trained to identify, analyze, and solve system problems and identify improvement opportunities.

Tree Diagram: A management tool used for analysis and planning where issues, activities, or ideas are broken down until actionable items are identified. It is primarily used for planning purposes where actions are planned at several levels below the top level of detail. It can also be used with the Five Whys technique of cause-and-effect analysis.

u Chart: The more general form of the c chart. It is used for plotting data based on the number of non-conformities (defects) found in each sample of either fixed or varying size.

Utilization: A measure of performance that relates actual activity performed in a process with potential activity (e.g., actual time performing activity/total time available including idle time). Processes that produce outputs a high percentage of the time available are well utilized.

Variable Data: Data obtained from quantitative measurement (continuous interval data), expressed as a fraction or decimal, that reflects a continuous measurement relative to some scale. Unlike attribute data, which have discrete values, variable data can be measured to greater levels of accuracy using more accurate measuring equipment. Also referred to as continuous data.

Variation: A measure of change in a variable, function, or process that illustrates that no two items will be completely the same.

Velocity of a Process: The elapsed time from the beginning to the end of a process (or cycle time) and/or the rate of production or productivity of the process.

Weighted Voting: A voting tool used to help teams identify preferences for various alternatives in their attempt to select courses of action or priority items using consensus. When used correctly, voting helps to decide favored alternatives by promoting discussion of differences.

x-R Chart: A pair of charts used to plot individual data elements, x, and the range between data values, R. Also known as individuals or moving average Control Chart. Used primarily for variable data, but can also be used for attribute data when the average count of defects or defective items in each sample is greater than two.

$\overline{X}$-R (x-bar–R) Chart: A pair of Control Charts used to plot variable data sampled from a process: $\overline{X}$ to track the mean of data values and R to track the variability, or range, of data values.

APPENDIX B

REFERENCES AND RESOURCES

Readers seeking more in-depth coverage of quality management topics should consult the following reference materials. For a more detailed listing of resources, contact GOAL/QPC at (800) 643-4316 for *The Source*, which contains 2400 annotated literature entries and other Total Quality Management resources. This 1600-page reference document comes in three volumes and is updated annually.

Various quality magazines, such as *Quality Progress, Quality Digest, and Quality*, provide excellent reference materials. In addition to annual indices to articles, they often publish useful listings of resources. For example, the March issue of *Quality Progress* includes their annual listing of quality-related software.

General Reference

Collins, Brendan and Ernest Huge. *Management by Policy: How Companies Focus Their Total Quality Efforts to Achieve Competitive Advantage*. Milwaukee, WI: ASQC Quality Press, 1993. [Policy Deployment]

Deming, W. Edwards. *Out of the Crisis*. Cambridge, MA: MIT, Center for Advanced Engineering Study, 1982.

Feigenbaum, Armand V. *Total Quality Control*, Third Edition. New York, NY: McGraw-Hill, 1983.

Imai, Masaaki. *Kaizen: The Key to Japan's Competitive Success*. New York, NY: McGraw-Hill, 1986.

Juran, J. M., Editor-in-Chief and Frank M. Gryna. *Juran's Quality Control Handbook*, Fourth Edition. New York, NY: McGraw-Hill, 1988.

Schonberger, Richard J. *World Class Manufacturing: The Lessons of Simplicity Applied.* Cambridge, MA: Productivity Press.

Sherkenbach, William W. *The Deming Route to Quality and Productivity.* Rockville, MD: Mercury Press/Fairchild Publications, 1988.

Walton, Mary. *The Deming Management Method.* New York, NY: The Putnam Publishing Group, 1986.

Control Chart Forms

American Society of Quality Control (ASQC): (414) 272-8575.

Ford Motor Quality Publications, c/o edcor Data Services (810) 626-3077.

Customer Service/Satisfaction

Albrecht, Karl. *At America's Service.* Homewood, IL: Dow Jones-Irwin, 1988.

Carlzon, Jan. *Moments of Truth.* New York, NY: Harper & Row, 1987.

Garvin, David A. "Competing on the Eight Dimensions of Quality," *Harvard Business Review.* Boston, MA: Harvard Business School Publishing, November–December 1987.

Hayes, Bob E. *Measuring Customer Satisfaction: Development and Use of Questionnaires.* Milwaukee, WI: ASQC Quality Press, 1992.

Jacobson, Gary and John Hillkirk. *Xerox, American Samurai.* New York, NY: MacMillan, 1986.

Whitely, Richard C. *The Customer Driven Company: Moving from Talk to Action.* New York, NY: Addison-Wesley Publishing, 1991.

Zeithaml, V. A., A. Parasuraman, and L. L. Berry. *Delivering Quality Service: Balancing Customer Perception and Expectations.* New York, NY: The Free Press, 1990.

Zemke, Ron. *The Service Edge.* New York, NY: NAL Books, 1989.

Other Tools and Techniques

Akao, Yoji. *Quality Function Deployment: Integrating Customer Requirements into Product Design.* Cambridge, MA: Productivity Press, 1990.

Alreck, Pamela L. and Robert B. Settle. *The Survey Research Handbook.* Homewood, IL: Richard D. Irwin, 1985.

AT&T. *Benchmarking: Focus on World-Class Practices*. Indianapolis, IN: AT&T, 1992.

Brassard, Michael. *The Memory Jogger Plus +™*. Metheun, MA: GOAL/QPC, 1989.

Camp, Robert. C. *Benchmarking, The Search for Industry Best Practices that Lead to Superior Performance*. Milwaukee, WI: ASQC Quality Press, 1989.

Fink, Arlene and Jacqueline Kosecoff. *How to Conduct Surveys: A Step-by-Step Guide*. Thousand Oaks, CA: Sage Publications, 1985.

Gogg, Thomas J. and Jack R. A. Mott. *Improving Quality and Productivity with Simulation*. Palos Verde Estates, CA: JMI Consulting Group, 1992.

Ishikawa, Kaoru. *Guide to Quality Control*. Englewood Cliffs, NJ: Prentice Hall, 1985.

Kepner, Charles H. and Benjamin B. Tregoe. *The New Rational Manager*. Princeton, NJ: Princeton Research Press, 1981.

King, Bob. *Better Designs in Half the Time*. Metheun, MA: GOAL/QPC, 1987. [QFD]

Marsh, Stan, John W. Moran, Satoshi Nakui, and Glen D. Hoffherr. *Facilitating and Training in QFD*. Metheun, MA: GOAL/QPC, 1991.

Morgan, David. L., Editor. *Successful Focus Groups*. Thousand Oaks, CA: Sage Publications, 1993.

Scholtes, Peter R. *The Team Handbook*. Madison, WI: Joiner Associates, 1988.

Statistical Process Control

Amsden, Robert T., Howard Butler, and Davida Amsden. *SPC Simplified: Practical Steps to Quality*. White Plains, NY: Unipub, 1986.

Box, George E. P., William G. Hunter, and J. Stuart Hunter. *Statistics for Experimenters: An Introduction to Design Analysis, and Model Building*. New York, NY: John Wiley & Co., 1978.

Chambers, David and Donald J. Wheeler. *Understanding Statistical Process Control*. Knoxville, TN: SPC Press, 1986.

Farnum, Nicholas R. *Modern Statistical Quality Control and Improvement*. Belmont, CA: Duxbury Press, 1994.

Grant, Eugene L. and Richard S. Leavenworth. *Statistical Quality Control*. New York, NY: McGraw-Hill, 1988.

King, James R. *Frugal Sampling Schemes*. Tamworth, NH: Technical and Engineering Aids for Management, 1980. [Probability Plots]

Mendenhall, William, Lyman Ott, and Richard L. Scheaffer. *Elementary Survey Sampling*. Boston, MA: Duxbury Press, 1986.

Nelson, Wayne. "How to Analyze Data with Simple Plots," *Quality Progress*. January 1979: 7–12. [Probability Plots]

Peace, Glen Stuart. *Taguchi Methods: A Hands On Approach*. New York, NY: Addison-Wesley, 1993.

Pyzdek, Thomas. *Pyzdek's Guide to SPC, Volume Two: Fundamentals*. Milwaukee, WI: ASQC Press, 1989.

Pyzdek, Thomas. *Pyzdek's Guide to SPC, Volume One: Applications and Special Topics*. Milwaukee, WI: ASQC Press, 1992.

Pyzdek, Thomas. "Process Control for Short and Small Runs," *Quality Progress*. April 1993: 51-60.

Sower, Victor, Jaideep Motwani, and Michael J. Savoie. "Are Acceptance Sampling and SPC Complementary or Incompatible?" *Quality Progress*. September 1993: 85-89.

Taguchi, Genichi. *The System of Experimental Design: Engineering Methods to Optimize Quality and Minimize Cost*. White Plains, NY: Unipub, 1987.

Teamwork, Leadership, and Communication

Belasco, James A. *Teaching the Elephant to Dance: Empowering Change in Your Organization*. New York, NY: The Free Press, 1987.

Bennis, Warren. *On Becoming a Leader*. New York, NY: Addison-Wesley, 1989.

Byham, William C. *Zapp! The Lightning of Empowerment*. New York, NY: Harmony Books, 1988.

Doyle, Michael and David Straus. *How to Make Meetings Work*. New York, NY: Jove Books, 1982.

Goldratt, Eliyahu and Jeff Cox. *The Goal: A Process of Ongoing Improvement*, Second Revised Edition. Croton-on-Hudson, NY: North River Press, 1986.

Kayser, Thomas A. *Mining Group Gold*. El Segundo, CA: Serif Publishing, 1990.

Senge, Peter M. *The Fifth Discipline*. New York, NY: Bantam Books, 1990.

INDEX

DATE DUE		
FEB 19 '96		
SEP 30 '96		

CONCORDIA UNIVERSITY LIBRARY
2811 NE Holman St.
Portland, OR 97211